Building Design and Development in Hong Kong

City University of Hong Kong Press

First published 2003
Second printing 2009
Printed in Hong Kong

ISBN-10: 962-937-079-4
ISBN-13: 978-962-937-079-4

Published by
City University of Hong Kong Press
Tat Chee Avenue, Kowloon, Hong Kong

Internet: www.cityu.edu.hk/upress
E-mail: upress@cityu.edu.hk

In memory of

Professor Bill Lim

Contents

Part I
Property Planning

Part II
Design and Management

Part III
Construction Process

Detailed Chapter Contents

Part I
Property Planning

Part II
Design and Management

Part III
Construction Process

Foreword

Raymond HO Chung-tai

The Division of Building Science and Technology of City University of Hong Kong has done a great job in putting together the divisional publication, *Building Design and Development in Hong Kong*. Characterized by its multi-disciplinary perspective, the book mirrors the diversified background and expertise of the Division's staff.

To students and readers in general, this book provides them with a broad picture of the building and construction industry in Hong Kong. To interested practitioners and earnest academics, it provides them with insights into selected topics. To all readers, the book is definitely a useful source of reference and information.

Ir Dr Hon Raymond HO Chung-tai, MBE, JP
Member (Engineering Functional
Constituency) of the Legislative Council
December 2002

Foreword

Patrick LAU

Building Design and Development in Hong Kong reflects the multi-disciplinary characteristic of the local property and construction industry. It provides a timely and valuable opportunity for us to share the most up-to-date information about a wide spectrum of inter-related topics that as a whole contributes to the design and development of buildings in Hong Kong. It represents a belief in the importance of collective effort and cooperation among all parties involved in the industry.

I recommend this book to all students and professionals in the field, as *Building Design and Development in Hong Kong* is an important source of reference for those who want to broaden their horizons and enrich their professional knowledge by crossing disciplinary borders.

Patrick LAU Sau-shing, S.B.S.
Professor, Department of Architecture,
The University of Hong Kong
President, Hong Kong Institute of Architects
December 2002

Foreword

LAU Ping-cheung

Vast contrasts exist in building design and development in Hong Kong. No visitors would not be fascinated by the uniqueness of Hong Kong's skyline, but locals are recently at odds as to whether planning restrictions should be imposed to control building heights. While new initiatives are being introduced to encourage more innovative and greener building designs, the restrictive and outdated Buildings Ordinance has been largely responsible for the monotonous and unsustainable building forms and shapes of our cityscape. Property owners should be responsible for the timely upkeep of their own buildings; yet the Administration is using public money through the Urban Renewal Authority to prevent and eliminate urban slumps as dilapidated buildings proliferate into a serious social problem.

The publication of *Building Design and Development in Hong Kong* at this critical juncture is most timely to enable readers to gain a broader view of the planning, design, construction and maintenance process in Hong Kong.

Hon LAU Ping-cheung
Member (Architectural, Surveying and
Planning Functional Constituency) of
the Legislative Council
December 2002

Foreword

Julie MO

With the population continuing to increase, the property market will remain a major driving force behind Hong Kong's economy. Although its contribution to economic growth may remain fluid for some time, property development demands our attention because the provision of a quality living and working environment is essential for Hong Kong to retain its position as an international metropolis, and to cater for the needs of a modern society in spite of competition for limited resources. It is in this context that the Division of Building Science and Technology came up with the idea of putting together a book on building design and development, which draws upon the knowledge our colleagues would like to share with those who are interested or involved in this industry, be they students, professionals, academics, or practitioners.

Although the Division of Building Science and Technology is a young department of only five years old, it has a relatively long history of evolution that dates back to the establishment of the City University of Hong Kong in 1984. The Division is committed to excellence in teaching and is actively involved in the society at large through consultancies, applied research and development, as well as professional activities and community services. It has benefited from the diversity in expertise and experience of a multi-disciplinary team of planners, architects, engineers, surveyors, and builders.

By tapping this rich pool of knowledge, the Division has been able to compile this comprehensive book. It covers a wide range of topics across the whole process of property development, from inception to completion, including territorial planning, building design, submission procedures, tendering, construction and more. The book is intended to be informative as well as forward looking; the authors have not only introduced the general principles of their respective fields, but also addressed Hong Kong's unique situation and the latest developments. I hope that readers will find this book useful and will enjoy reading it.

Julie MO Kwok-wah
Head, Division of Building Science and Technology,
City University of Hong Kong
December 2002

Preface

Ellen LEE

The work has been the effort of a group of lecturers of the Division of Building Science & Technology who teach different subjects. The theme of the book is to cover multi-disciplinary expertise of the Division and is therefore focused on the local practice of the industry. Although some chapters provide certain theoretical principles, they are also written in a way comprehensive at practice level and are intended to give the readers a broad view of the design and construction process in Hong Kong.

Synopsis of the book

The book is intended to reflect the building design and property development practice in Hong Kong where relevant publications are in short supply. It is focused on three areas: property planning, design and management, and construction and maintenance.

Part I   Property Planning

Chapter 1 starts from the historical planning practice and the planning strategy developed today, highlighting the Territorial Development Strategy that has a great impact on land use development for the last 10 to 20 years. Chapter 2 leads us to the development of design, a stage to be carried out after planning approval has been given. This stage is where the Building Ordinance takes effect on design, including the different Building Regulations and Codes of Practice in use that significantly affect the design submission and approval process. It leads us to understand the function of the Building Authority, where safety and health acts also have an impact on design and construction of building works. Besides affecting the design of new buildings, the Buildings Ordinance also affects existing buildings. Chapter 3 describes a common problem in Hong Kong: unauthorized building works. It highlights the government's policy on unauthorized building works.

Part II   Design and Management

Chapter 4 departs from government policies and begins a journey of "finding form" (i.e. how space is conceptualized in an architectural design) in the early stages of the building design process. In chapter 5 the author traces the changes in architectural design in Hong Kong from the early decades of the 20th century to the present, describing how Hong Kong has evolved from a British colony into an international metropolitan in southern China. Chapter 6 provides practical examples of spreadsheet applications for the design and analysis of structures of tall buildings in Hong Kong. Chapter 7 extends the computer application to be used as a simulation tool to building services design. Chapter 8 describes the tendering systems in use to select contractors to carry out building works by public clients and private clients in Hong Kong.

Part III   Construction Process

Chapter 9 discusses the factors to be considered in site layout planning, using a real-life case as illustration. Chapter 10 examines the considerations for temporary works in building construction with an emphasis on safety issues. Chapters 11 and 12 describe the formwork systems and the technology used for complex high-rise buildings with illustrative cases. Finally, Chapter 13 highlights the effects of building services on construction processes and emphasizes the importance of effective coordination.

This publication is part of the work of the Applied Work Group led by Dr. T.T. Chow. The major issue of getting a book like this ready for publishing is coordination and quality assurance, which is the work of the Publication Work Group. While having Professor Bill Lim acting as our external reviewer to oversee the development of this book, we have also invited local reviewers to comment on the manuscipts. I would like to thank the reviewers for their generous help and advice:

Mr. Edwin H W Chan, Department of Building and Real Estate, Hong Kong Polytechnic University
Mr. Ben Chong Wing Hong, Building Department, HKSAR
Mr. Francis Leung, Architectural Services Department, HKSAR
Mr. Kong Man, Ove Arup & Partners Hong Kong Ltd

In addition, the book will not be in place without the effort of Charlie, Loonie, Raymond and Paul of the Work Group, the help of Sarah from our general office, the small meetings and the big meetings in between, and the support shown by the City University Press, Mr. Patrick Kwong in particular.

Ellen LEE
Chairman of Publication Work Group
December 2002

Prologue

Bill LIM

Education and training of technologists for the many professions in the building industry is as old as the industry itself. Training of craft skills helped maintain sufficient numbers of craftsmen in ancient Egypt and Babylon, and teaching the young was required by the laws of Hammurabi as early as in the 18th century BC. Standards of trades were kept by the Roman *Collegia*, and by the 13th century craft guilds were established to supervise quality and methods of production and regulated conditions of employment. Master craftsmen recruited apprentices who entered into a period of training, usually as long as seven years. The medieval guild made its members responsible for the competence of their apprentices.

The demand for technologists since the Industrial Revolution brought about mechanics' institutions including the Polytechnic (*Ecole Polytechnique*) in France in 1794, the Birkbeck College in London founded by George Birkbeck in 1823, and the Cooper Union for the Advancement of Science and Art in New York City in 1859. The French institution was the first of the great professional schools (*grandes ecoles*) with many disciplines.

Modern universities were developed from the medieval schools known as *studia generalia*, recognized as places of study open without restriction to students from all parts of Europe, formed as private societies for the personal interests of their members, and by the 14th century they became communities of scholars and teachers.

The polytechnics developed in the United Kingdom and adopted elsewhere responded to the need of science and technology education especially in the post-war years apart from the universities, with emphasis on providing competent technologists who could be readily absorbed into the rapidly expanding industries. Over time the polytechnics have become degree-conferring institutions with little distinction from universities, and in due course some have established non-technological programmes.

It is within this context of technological education that the Division of Building Science and Technology of the City University of Hong Kong produced this publication. Formerly the City Polytechnic of Hong Kong, the University has inherited the European and British tradition of technological education, with strong emphasis on practicality and realism. Since the establishment of the Hong Hong Special Administrative Region of the People's Republic of China in 1997, the Division is in a unique position in the education of building technologists who are able to meet the challenges not only in the former colony but also in a country which has about a quarter of the total population of the world.

The Division continues to fulfil its mission as part of the University in meeting its obligations to the needs of the building industry as stated in its mission statement:

> (to) provide quality higher vocational education for students to meet changing needs of the building industry;
> enhance the learning skills and encourage the continuous development in personal potentials of students;
> contribute to the advancement of knowledge in the building industry through applied research and consultancy; and
> reach out to the community by servicing, interacting and cooperating with other institutions of higher education, building-related professional bodies, government organizations, employers, and the building industry.

This publication specifically aims to reach out to the community to inform, inculcate and inspire through a series of essays written by staff of the Division. The theme of the publication, as set forth by the Publication Work Group set up in 1999/2000, is to present the multi-disciplinary expertise of the Division and demonstrate its relevance to the building industry.

The publication firstly serves as a channel of information and communication for the building industry in Hong Kong. It presents a broad view of the operation of the industry, its structure, process, and development, including the inter-relationship of the many professionals engaged in the day-to-day working environment. It also attempts to clarify and to differentiate the professional knowledge, skills, and specialization of the various technologists trained in the Division. This helps the public to identify and to engage the appropriate personnel for specific tasks within the building the industry.

Such information is also for students who are studying the programmes offered by the Division, or for potential students who aspire to work in the industry, to understand the multiplicity and the complexity of the building industry so that they are able to identify their own positions in the industry and to relate to others. Lack of understanding of other professions' roles often leads to ignorance and even prejudices, and this publication is expected to minimize some of the uncertainty in the minds of students concerning contributions by all professions and to encourage good working relationship both as students and as graduates. Thus, this publication is a useful reference for the different disciplines within the Division, though it is not intended to be a textbook.

The second aim of the publication is to inculcate in the academic and practice communities, both in Hong Kong and abroad, a sense of camaraderie and fellowship. Within the academic circle, opportunities to exchange ideas and to conduct multi-disciplinary teaching and research can only be enhanced if free flow of information and communication is available. This publication presents the work of the Division comprehensively as an introduction to the academic community so as to generate stronger linkage and cooperation.

The practice communities have a long-time association with the University in accreditation and collaboration in research, consultancy, and life-long learning activities. This publication gives the practice communities an overview of the interests and capabilities of the Division's staff, and this comprehensive reference helps to foster and to strengthen relationships.

The third aim of the publication is to inspire students, graduates and professionals in their quests for appropriate and innovative responses to the many issues confronting the building industry to date. Students are urged to develop self-learning and problem-solving skills, and become less dependent on lecturers, and therefore more self-reliant in the teaching and learning process. This is important in the newly generated associate degree environment in which students are considered to be mature enough to manage undergraduate studies. The cultivation of aptitudes and competence required in the building industry is paramount. Students are also encouraged to develop the appropriate knowledge and skills which will enable them to enter the workforce with confidence.

As graduates acquire generic as well as professional skills to meet the challenges of the professions, they need to think laterally and to work intelligently. Ethical work habits and leadership qualities are determining factors for success. While the two short years of associate degree programmes introduced in 2000 are considered to be sufficient for the commencement of professional life, the development of the full potential of the graduates depends on their attitude and commitment to life-long learning. Without consistent acquisition of knowledge and up-dating of skills they are not only left behind, but may even be replaced by a newer generation of professionals.

The advancement of building technologies brings as many problems as answers, and astute discernment is required of the professional before the adaptation of a new technology under prevalent local conditions. Sustainability in all aspects, including politics, economics, society, culture, and technology, is to be the key issue in the development of the built environment, and it must be considered holistically. Reforms in the legal and professional aspects of the industry are to be reviewed so that they may respond to the ever-changing environments in the industry.

To respond to these issues the staff of the Division have written papers in their respective fields of interests and expertise. The papers address specific issues of education and practice relevant to particular sectors of the building industry. Some comment on legislation and control; others summarize application and practice. Constructive criticism and reflections on current developments are given, and innovative solutions to problems are introduced.

Being a channel of information and communication for the building industry, this publication is intended to be used by senior students and graduates as well as young professionals who are adapting themselves to new areas which are not within their institutional training; it will also broaden their knowledge to a fuller understanding of the building industry.

It is the fervent hope of the Division that, in presenting this publication to academics and practitioners in the building industry, there will be a concerted effort from all concerned to bring about excellence in education and advancement in building science and technology. It also looks forward to further collaboration among the professional community in the future.

Bill LIM
(Former) Emeritus Professor,
Queensland University of Technology, Australia
December 2002

List of Illustrations

Figures

Tables

Part I
Property Planning

Chapter 1

Planning, Territorial Development Strategy (TDS) and Development Control

Kevin MANUEL

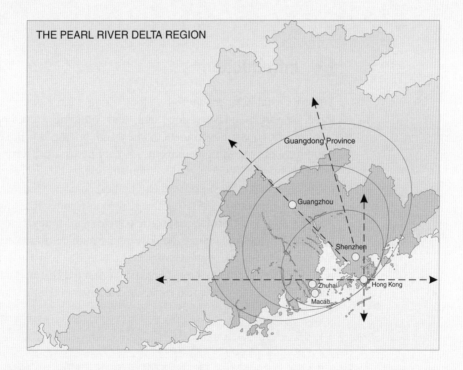

THE PEARL RIVER DELTA REGION

Hong Kong is facing great challenges from its neighbouring cities. One way of maintaining Hong Kong's competitive edge is to reshape the built environment by means of good city planning. This chapter introduces governmental strategies in territorial development and outlines the future territorial development of Hong Kong in a macro prespective through examining the background, process and mechanism of TDS, as well as overall land use planning. It also attempts to illustrate the relationship between TDS and development control.

1

Planning, Territorial Development Strategy (TDS) and Development Control

1 Introduction

Since its establishment in 1841, Hong Kong[1] has relied heavily upon three important players for land and building development: the government, quasi-government bodies and private developers. In the building industry, the government, also acting as a property developer, provides building control, land administration, public infrastructure and public housing. Quasi-government bodies, such as the Mass Transit Railway Corporation (MTRC) and Kowloon-Canton Railway Corporation (KCRC), may formulate specific policies and strategies together with the government in planning and developing public infrastructure such as the Mass Transit Railway network and the KCRC's West Rail and East Rail. Private developers have played a substantial role and carried out a substantial portion of the real estate development in Hong Kong.

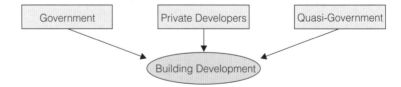

Figure1.1   Participants in building development

The over-concentration of resources in land matters and administrative duties on building and infrastructural development has enabled the government to undertake massive building and infrastructural works in Hong Kong in the previous decades. Nevertheless, due to the limited land supply, land value had been consistently at a high level prior to the financial crisis in 1997 and 1998. The lack of comprehensive planning[2] up to the 1970s had caused severe environmental and economic problems; for example, inadequate transportation networks in newly developed and developing areas, incompatible land uses, high cost of housing, inadequate amenities and facilities, environmental deterioration, and the loss of historical heritage public space. Hong Kong currently faces a dilemma between harbour reclamation and the extremely high density of development that has led to social and environmental problems.

2   Objectives

It is necessary for students, building professionals and the general public to appreciate the nature of planning, the Territorial Development Strategy (TDS) and the background of planning in Hong Kong. The main objectives of this chapter are mainly threefold. First of all, it is to introduce to students and laypersons the background of overall town planning in Hong Kong and some major issues related to the TDS. Secondly, it is to examine the background, process and impact of the TDS, and overall planning on our built environment. Thirdly, it is an attempt to examine some of the key issues of territorial planning and to give comments and recommendations for further discussion and action. This chapter will give an overview of the TDS and its impact on building development in Hong Kong.

3   What Is Planning and Its Purposes?

Planning allows development of the right land use at the right time and in the right place. Planning in the context of this chapter refers to land use planning. Town, city or urban planning in Hong Kong usually means land use or physical planning. The scope of planning embraces the physical, social and econoimic aspects of land use. These aspects intersect each other and is illustrated in Figure 1.2.

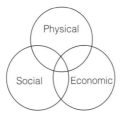

Figure 1.2   Scope of planning

Another view on planning is its three dimensional charateristics. One deals with different parties and interested groups; one is about the scope of space and land use; the other is concerned with the dimension of time, and the social and political context.

Prior to the 1980s, town planning in Hong Kong adopted an engineering and mechanistic approach with roads and functional land uses and other kinds of infrastructural support forming a basic formula. Architects, landscape architects, urban designers and the general public were usually not involved in the early stages of planning. As a result, social considerations were not mature and environmental considerations with regard to sustainability[3] were a myth. The situation gradually changed as the former government decided to build a new airport in 1989 owing to overcrowding conditions at the former Kai Tak Airport. This major move marked the beginning of the ten core infrastructure projects, which required a much more integrated and well-planned strategy. It also marked a new era of town planning in terms of a long-term development strategy[4] forming a broad-based framework for future development.

The hierarchy of planning is illustrated in Figure 1.3. It includes three tiers: general-strategic, subregional, and local and site-specific. The principles and objectives of the top tier — the TDS covers the long-term broad and strategic framework that is gradually refined at subsequent tiers.

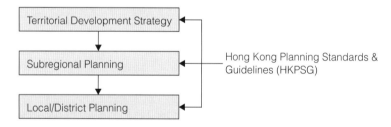

Figure 1.3   Hierarchy of plan

4   Major Events of Town Planning in Hong Kong

Planning and its control prior to 1939 were only enforceable by a few piecemeal ordinances, such as the Building Ordinance of 1889, the Close House and Insanity Dwellings Ordinance and the Public Health and Building Ordinance of 1903.[5] The milestone in land use planning in Hong Kong was the enactment of the Town Planning Ordinance (TPO) in 1939 "to promote the health, safety, convenience and general welfare of the community". Under the Ordinance, the Town Planning Board (TPB) was appointed to prepare draft plans for the future layout of existing and potential urban areas. However, there was *little enforcement* and no implementation strategy at the time. The Second World War caused a halt in planning and building activities, and no planning was made to achieve a high quality living environment.

In 1948, Sir Abercombie, a British town planner, completed a report on guidelines and future planning for the urban areas and ports of Hong Kong. The report proposed a cross-harbour tunnel, reclamation along the harbour, housing supply, new town development and relocation of military barracks, etc. Many of these proposals were adopted.

In 1953, a Planning Branch was established in the Public Works Department. Emphasis was placed on the preparation of statutory plans under the auspices of the Town Planning Board. The first Outline Development Plans for Shatin was approved and published in 1962. Ideas were also put forward to establish country parks for recreation and conservation purposes. In 1965, the government initiated the Colony Outline Plan (COP) to review future land use and the population distribution pattern throughout the territory. In 1970 the COP was completed with recommendations on the comprehensive planning of land use and the population distribution pattern throughout the territory. The COP was approved by the Executive Council (ExCo) in 1972.

In 1972, the "Ten-Year Housing Target Programme" was announced by Sir Mclehose to develop the various new towns in the New Territories, such as Tuen Mun, Shatin and Tai Po. Since then town planning in Hong Kong entered a new era, aimed at housing another 1.8 million people by the mid-1980s. By the late 1970s, many of the "first generation" new towns were heading towards maturity. While the search for development space continued with the development of the "second generation" new towns, there was a rising concern for the need to coordinate development efforts. In addition, there was also a concern over the congested urban structure in the main urban areas and the increase in demand for new ports, airport facilities and land-based transportation infrastructure. In 1974, the COP was substantially revised to take account of the changes in socio-economic conditions and the government's development priorities. The revised plan, including a set of planning standards and a development strategy, was re-titled as the Hong Kong Outline Plan (HKOP).

In the 1980s, the planning profession in Hong Kong began to place more emphasis on territorial and sub-regional planning. The government refocused on the main urban areas with renewed interest. The Metroplan, as a major plan at the sub-regional level (Figure 1.4), was one of the most important documents to examine the issues and potentials in the metro area. For the older urban areas, apart from continuing redevelopment and environmental improvement, an urban renewal strategy was completed in early 2000 to enhance the redevelopment potential of urban areas. However, not much effort with respect to sustainability was put into the formulation of the TDS until the middle of the 1990s. Planning activities were not very enforceable at the local and district level until the Amendment of the TPO in 1991. In sum, the planning intentions of the TDS are implemented through local plans at district levels but enforceable through the Buildings (Planning) Regulations.

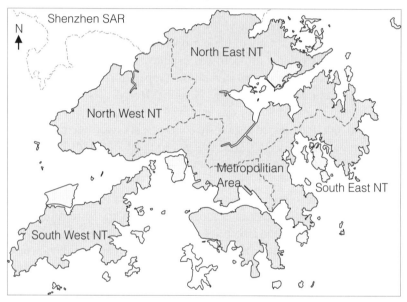

Figure 1.4   Hong Kong and its 5 main subregions

5   A Glance at the Planning and Development Process (Private Sector)

A building development is mainly under the control of three key jurisdictions, namely lease, the town planning ordinance and building ordinances, along with other related ordinances under the control of various authorized departments. Land use is designated under the lease conditions and Outlined Zoning Plans (OZPs) in urban areas or the Development Permissions Area Plans in rural areas.[6] These plans are prepared at the local and district levels but with the planning intentions and considerations of the subregional plans and the TDS. A building design proposal is approved if it is in compliance with the regulatory aspects of the development control process listed in Figure 1.5.

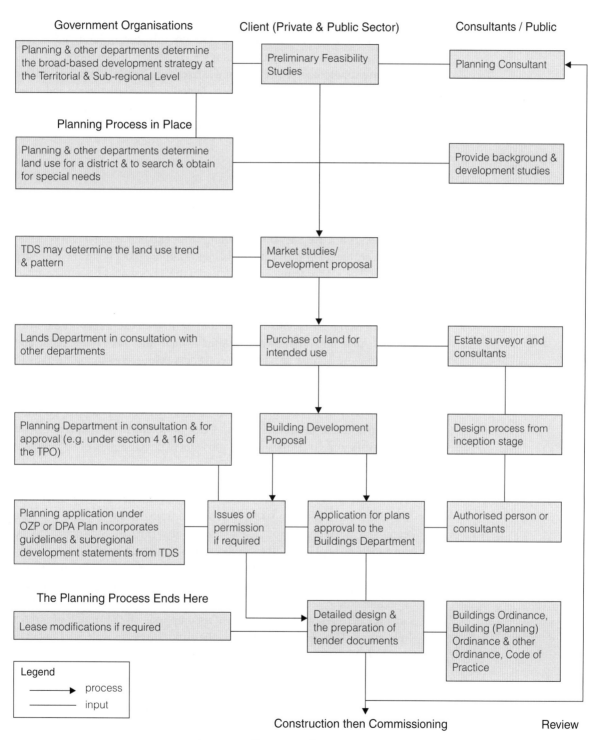

Figure 1.5   The planning and building development process in Hong Kong

Development proposals are examined and approved by the Buildings Department as a central agency and coordinator of comments and approvals by various departments under different jurisdictions, such as fire, environmental protection and lease conditions. Planning policy, strategy and intentions can only be reflected in the district plans under the jurisdiction of the TPO. The TPB, according to the power vested under the TPO, reviews, examines, and make decisions on land uses that contravenes the statutory plans.

6 The TDS, Its Functions and Background

The TDS is not a legally binding framework and hence is not enforceable. However, the TDS provides the background and backbone for planning studies. The planning intentions stated in the TDS would be implemented and reflected at the district level through local plans. The TDS is also a kind of planning activity initiated and executed by the planning authorities in Hong Kong "to establish a broad long-term land use-transport-environmental planning framework within which the necessary land and infrastructure can be provided. It considers resource availability so as to enable Hong Kong to continue to grow as a regional and an international city and become a better place in which to live and work"(TDSR 1996). In order to provide an orderly, compatible and efficient land use for building and infrastructural development, TDS is established to permit such framework and guidelines for the subsequent development of subregional planning and district planning.

The TDS was first formulated by the government in 1980 and was completed in 1984. It provided an overall framework for territorial development, mainly in terms of land and transportation use to promote economic development of the main urban area (metro). Subsequent to the formulation of TDS-1, the Strategy was reviewed in 1986 using new demographic forecasts and the same set of assumptions on airport and port developments. The results of the review, TDS-2, generally confirmed the previous studies and provided a similar set of predictions. In 1988, TDS-2 was further reviewed with different assumptions on the timing of airport relocation. TDS-3 of 1988 assumed the airport to be relocated to Lantau in 2001. TDS-4 extended the study to 2011 and considered a wider region covering southeast China. Basically, the focus of the strategy was placed on the urban areas and the reason was partly twofold: first, to make the urban areas more suitable for transportation and economic development and second to redevelop the decaying urban areas while considering their potential after the relocation of Kai Tak Airport.

The TDS Final Executive Report was approved by the Executive Council in February 1998. In addition to recommendations on housing, employment, transport and the environment, the TDS review also looked at "distant vision concepts" for further consideration in subsequent revisions. Since 1997, studies have been launched under the government's Sustainable Development for the 21st Century (SUSDEV2)[7] iniative and many studies are being considered in consultation with the public and professional bodies. Hence, the TDS dictates the growth pattern and phases of development, distribution and allocation of land use.

7   The TDS and Development Control at Subregional and Local Levels

Development as defined in the Town Planning Ordinance (TPO) is comprehensive and covers a diverse range of human activities. In Hong Kong, development control[8] is mainly monitored at the district level even though the scope could be inter-regional or territory-wide, as in the case of the Hong Kong Airport (Control of Obstructions) Ordinance and the "Ten Core Projects" of the new airport development in the 1990s. Hence TDS can have a significant impact in three aspects. First of all, it may help provide broad planning intentions and guidelines and guide decisions at the district and subregional levels. As the strategies and planning goals and intentions are filtered down to the district levels, they may form part of the planning intentions of the outlined zoning plans (OZP) and development permission area plans (DPA). However, the scope of development control is restricted to areas covered by DPA plans and areas where OZPs in the rural areas of the New Territories have replaced DPA plans. These plans generally incorporate the planning goals and intentions of the TDS and provide guidance for planning control.

Secondly, the TDS can be further developed into a number of subregional development statements (SRDS) and development strategies (SRDSG). These documents and reports are mainly the products of internal government cooperation and collaborations. Lastly and most importantly, the TDS can be channelled through to other departments or Bureaux via inter-departmental communication. The indirect impact of TDS may be crucial as it embraces environmental, social and economic goals or objectives. If other departments implementing certain Ordinances, such as the Environmental Protection Ordinance and the Environmental Impact Assessment Ordinance (EIAO), also incorporate the essence of the TDS "spirit", its impact on development control decisions can be substantial, as in the case of the Lok Ma Chau Spur Line of the East Rail and the Lantau North South Road Link. This indirect impact has not been researched but the effect could be significant if members of the Territorial Development Strategy Steering Committee (TDSSC) and Bureaux could channel the TDS principles and strategic guidance clearly to the individual departments or offices.

Although the TDS provided a broad framework to guide the overall development of Hong Kong since the 80s, it does not have any statutory power. However, the TDS does provide an integrative approach to territorial development matters. These include the distribution of transportation network, strategies for conservation, the adequacy of land for housing and related infrastructural development, and other provisions of services and support for economic and social activities that will yield environmental benefits.

In principle, development control is only effective when imposed under the Buildings Ordinance, the leases and some other important ordinances, such as the EIAO, Antiquity and Monuments Ordinance (AMO) and Road Traffic Ordinance (RTO). Other administrative measures such as Special Control Areas, Sites of Special Scientific Interests and Density Zoning also have a great impact on development control. The simplified process of building development from land acquisition to planning approval is shown in Figure 1.5. In brief, the direct impact of TDS on day-to-day operations is only minimal because it does not have any statutory power. However, if members of the TPB integrate the planning principles and intentions of the TDS into subregional and district levels, the TDS can have a more direct and comprehensive, since it also provides important guidelines and principles for decision making at the local level.

8 The Outlined Process of TDS Formulation and a TDS Case Study

Traditionally, TDS adopts a typical process of the rational comprehensive planning model (Chapin & Kaiser 1979, Freidmann 1987). In principle, the model consists of nine major key activities as illustrated in Figure 1.6. For a complex issue such as the development of Chek Lap Kok Airport, the actual implementation of the process can take up to ten years.

Back in 1989, the Hong Kong Government decided to relocate the airport to Chek Lap Kok and to expand port facilities in North Lantau and Western Harbour area. In light of the rapidly evolving China-Hong Kong economic linkage, a review of the TDS becomes necessary to take into account the possible consequences of the new plans. Subsequently, the

Planning Department published a TDS Review Consultative Document in 1993. This review aims to establish a broad planning framework for the preparation of subregional and district plans as well as the integration of public policies on land and infrastructure development. In an effort to keep pace with development trends in South China, particularly the Pearl River Delta, the TDS was critically reviewed. The ten core project proposals such as the Tung Chung Development (Phase I), West Kowloon Reclamation and Central and Wanchai Reclamation were drawn up to work along with the TDS framework to provide efficient and feasible solutions to development needs.

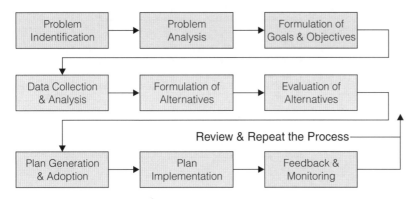

Figure 1.6   A typical planning model

The TDS review re-examined several major issues relating to land use development in Hong Kong, including future population growth, port and airport development, economic activities, current policies on land uses, environmental quality and transport systems. During the review, the public was consulted in 1993 and 1996.

9   The Planning Mechanism

Figure 1.7 illustrates the planning system of the three tiers and indicates whether statutory power is vested at these three levels. In terms of strategy formation, it may work both ways from the departmental level up and from the bureau level down, as illustrated in Figure 1.8. The policy determined by senior government officials[9] would spell out the emphasis and direction of TDS, as seen from the 1980s–1990s. The improvement and success of current integrative transportation network had been the important direct and indirect result of the well-planned strategies by the gvernment in the 70s and 80s. Nevertheless, economic development was the top priority in the 1980s. Hence, the emphasis on land use and planning was mainly on efficiency and economic and industrial strategies.

The TDS is a set of very comprehensive programmes developed over many years of research and investigation. It is often anticipated that during the process of study, new parameters and factors will be incorporated for continuous improvement and revision.

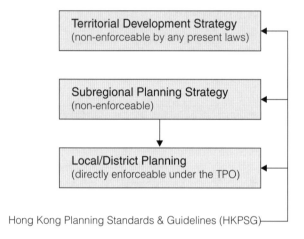

Figure 1.7   The planning system

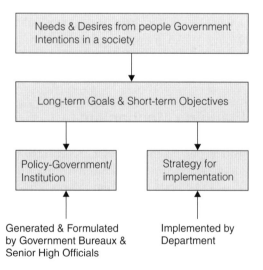

Figure 1.8   Strategy formation

10 Impact of the TDS: 1980s to 2001

The TDS was first completed in 1984. Then came various strategic decisions made by the Government on the Port and Airport Development Strategy (PADS) and transport and environment in 1989, and later on Metroplan[10] in 1991. These decisions have a long-term impact on the future urban fabric. The priority was to address interface issues arising from the above strategic decisions. The strategic development implications of the relationship between Hong Kong and South China have been addressed at the TDS level since the late 1980s. However, at the subregional and district levels it had not been able to incorporate many key environmental and living quality issues, such as urban design and renewals and conservation strategy of the natural environment.

The set of principal objectives derived from the key TDS reports prior to 1996 was related to economic, social and environmental development, such as the enhancement of Hong Kong's role as an international financial city, and ensuring sufficient provisions to be made for land use and infrastructural needs. In this connection, land has always been considered a commodity for stimulating economical growth.

Beginning from the 1990s, the TDS has been focusing on the importance of Strategic Growth Areas (SGAs) in the New Territories, infrastructural and cross boundary issues and the whole Pearl River Delta area (PRDA). It further addressed the close relationship between mainland China and Hong Kong. In brief, the merits and shortcomings of the TDS are listed below:

a) It has been able to provide guidelines and directions for a more detailed planning at the sub-regional and district levels, e.g. providing a greater opportunity for developing and formulating a tourism strategy.

b) The concept is more of a "top down" type than a "bottom up" grass root type prior to the 1990s.

c) The TDS anticipates a positive economic growth at all times and does not make provisions for stagnant or negative economic growth.

d) The strategy has not yet been able to initiate a natural conservation policy, a marine conservation policy or a comprehensive landscape strategy at the sub-regional levels.

e) The TDS does not address the image of the city such as the protection of the views on both sides of the Victoria Harbour or obained clear policy support from senior Government officials, the Bureaux, ExCo and the politicians.

f) TDS is not able to address the whole spectrum of sustainability, especially environmental and social sustainability. TDS is not able to seek more public views in the form of public consultation and public exhibits around Hong Kong prior to 1990s. These are the failure of certain key strategies at the district and region levels, such as the Victoria Harbour Reclamation and SE Kowloon Development in particular.

g) TDS requires very close collaboration with various policy branches and government department in providing necessary input on policy issues and assumptions. However TDS/Steering Group, usually does not indicate a clear set of well-consulted and accepted policies backed up by various Bureaux, backing up by senior high government officials to provide guidance and direction on major studies and to monitor the general progress of the study.

h) The study should allow opportunity for carrying out detail and local urban design and environmental improvement and not to focus on land use matters alone.

In February 2001, the Hong Kong Government launched the Hongkong 2030 campaign, which is a courageous, far-sighted attempt to predict the "idealized" or "ought to be" conditions of Hong Kong in the next 30 years or so. Many people would argue that TDS should not cover such a long time frame. The duration of the study in fact is not the key issue. The key to the success of any strategy is the government's financial, administrative and regulatory support. Therefore, the TDS should be backed by a clear and strong policy statement from the Bureau and be incorporated in the TDS. The "doing just enough" approach by the senior government officials should be abandoned. Indeed, in the past the government and some elite groups in the society have discouraged many innovative and courageous attempts to reform the planning system and policy and implementation framework, even though the problems arising from many policies and strategic plans had been thoroughly discussed and studies conducted. Due to changes in the economic and social conditions, many policies on natural conservation and environmental protection were not carried out.

A Stage 1 Public Consultation was held in February and March of 2001. A few thoughts about how to make Hong Kong more sustainable were elaborated at the subregional level. Hopefully, the later stages of the TDS would reveal more concrete details of strategic implementation. In sum, Hong Kong 2030 has many merits, some of which are listed below:

a) a more realistic approach towards determining the strengths and weaknesses of Hong Kong as an international and world-class city as well as the opportunities and challenges ahead;

b) a vision and strategy that incorporates the issues of sustainability;

c) public consultation right in the beginning;

d) a deeper study of cross-boundary issues and the Pearl River Delta Area;

e) adaptive short-term plans and flexible long-term plans catered for simultaneously;

f) better support by senior government officials with the integration of the Decision Support Tool developed through the SUSDEV21.

11   Conclusion

To conclude, the TDS provides the steering strategies and framework for the allocation of land resources for development. It is geared towards all kinds of development including building and infrastucture. Its development is becoming much more open and transparent. Several key considerations are required to make the TDS more influential. It may incorporate the concepts of sustainability with the aid of the Decision Support Tool (CASET) (Planning Department and ERM 2000) so that key strategic decisions — concerning its formulation can be made based upon a set of scientific and analytical tools. In turn, hopefully, the Government decision making process will facilitate TDS development and overall planning with respect to social, economic and environmental considerations. The broader and deeper the study of Hong Kong 2030, the better the prediction and planning for the common future of Hong Kong. More developers and design and planning professionals now realize the importance of the TDS because it affects development and investment decisions. The land reserve pattern of developers would then reflect the growth patterns as indicated in the TDS. Having read this chapter, the reader should notice the impact of the TDS on devlopment control and the long-term impact of TDS on Hong Kong's planning development.

Notes

1. Since July 1, 1997, Hong Kong's official name has become "Hong Kong Special Administrative Region".

2. Planning — It refers to study, planning, evaluation, implementation, and enforcement of land use resources and development. It includes land use planning and physical planning, transportation planning, country or rural planning, social planning and economic planning.

3. Sustainability is defined as "development which meets the needs of the present without comprising the ability of future generations to meet their needs" in the World Commission Report, 1987. The three categories, social, environmental and economic, are also the three basic components of sustainability.

4. A strategy is usually developed under a specific policy guideline, statement or objective, or a goal developed under a specific ordinance or administrative statement. As a strategy is formulated for implementation and consultation whereas a policy is developed for consultation and intra-government discussions. A policy may not lead to action and implementation.

5. Refer to R. Bristow's *Land Use Planning in Hong Kong*, and V. Lampugnani's "Aesthetics of Density" for further reading.

6. Both OZP and DPAs are Statutory Plans, prepared with the factors and land use patterns considered by the TDS.

7. It was a study carried out by PlanD and ERM Ltd. from 1997 to 2000 to develop a systematic process to enable decision makers to understand the long-term implications of development by the use of a set of sustainable indicators.

8. Development control is defined as a mechanism in which the Government uses a set of regulatory or administrative measures to guide and control private development.

9. They refer to the secretaries and their assistants, ExCo members & the Chief Executive.

10. It is a type of sub-regional planning focusing on the main urban areas (metropolitan area) of the city (refer to Map 5.4, "Town Planning in Hong Kong, 1988).

References

1. Bristow, Roger. 1987. *Land Use Planning in Hong Kong: History, Policies and Procedures*. Oxford: Oxford University Press.

2. _____. 1989. *Hong Kong's New Towns*. Oxford: Oxford University Press.

3. Chapin, F. Stuart & Kaiser, Edward. 1979. *Urban Land Use Planning*. Chicago: University of Illinois Press.

4. Freidmann, John. 1987. *Planning in the Public Domain: from Knowledge to Action*. Princeton: Princeton University Press.

5. Fung, Bosco C. K. 1988. "Enforcement of Planning Controls in Hong Kong." *Planning and Development* (Journal of the Hong Kong Institute of Planners). 4 (1): 21–26.

6. Hong Kong SAR Government. 1990. Consultative Document: Interim Amendments to the Town Planning Ordinance. Hong Kong Special Administrative Region.

7. Hong Kong Institute of Planners. 1986. *Planning and Development* (Journal of the Hong Kong Institute of Planners). Various issues.

8. Hong Kong SAR Government. 1991. *Comprehensive Review of the Town Planning Ordinance Review*. Hong Kong Special Administrative Region.

9. _____. 1995. *Town Planning in Hong Kong*. Hong Kong Special Administrative Region.

10. Lai, Lawrence, W. C. 1997. *Town Planning in Hong Kong: A Critical Review*. Hong Kong: City University of Hong Kong Press.

11. Lampugnani, V. 1993. *Aesthetics of Density*. New York: Prestel-Verlag.

12. Lo, W. M. & Chan, L. T. 1999. *Introduction to Town Planning in Hong Kong*. Hong Kong: Joint Publishing.

13. Planning Department. 1996. *A Consultative Digest of Territorial Development Strategy*. Hong Kong Special Administrative Region.

14. _____. 2001. *Hong Kong 2030, Planning Vision and Strategy, Stage 1 Public Consultation*. Hong Kong Special Administrative Region.

15. _____. *Hong Kong Planning Standards and Guidelines: V.1–V.11*. Hong Kong Special Administrative Region.

16. _____. 1998. *TDSR-A Response to Change and Challenges, Final Executive Report*. Hong Kong Special Administrative Region.

17. _____. 1996. *TDSR-96, Report on Public Consultation*. Hong Kong Special Administrative Region.

18. _____. 1998. *Port Development Strategy, 1998 Review*. Hong Kong Special Administrative Region.

19. Planning Department & ERM. 2000. *Sustainable Development for the 21st Century-Executive Summary*. Hong Kong Special Administrative Region.

20. Planning Environment and Lands Branch, Strategic Planning Unit. 1990. *Metroplan: The Aims*. Hong Kong Special Administrative Region.

21. _____. 1990. *Metroplan: Initial Options*. Hong Kong Special Administrative Region.

22. _____. 1990. *Metroplan: The Selected Strategy, An Overview*. Hong Kong Special Administrative Region.

23. _____. 1990. *Metroplan: The Foundation and Framework*. Hong Kong Special Administrative Region.

24. Pun, K. S. 1988. "Planning Achievements in Hong Kong since 1953." Arthur Ling, ed., *Urban and Regional Planning and Development in the Commonwealth*. Howell Publications. 217–222.

25. _____. 1989. "Past and Future Development of Urban Planning in Hong Kong." *Planning and Development* 5 (1): 7–13.

26. _____. 1984. "Urban Planning in Hong Kong — Its Evolution Since 1948." *Third World Planning Review* 6 (1): 61–78.

27. Town Planning Office. 1989. *Town Planning in Hong Kong*. Hong Kong Special Administrative Region.

28. World Commission on Environment and Development. 1987. *Our Common Future, The Report of the World Commission on Environment and Development (The Brundtland Commission)*. Oxford: Oxford University Press.

Statutes

1. Town Planning Ordinance (U.K. 1991)

2. Hong Kong Town Planning Ordinance (1999)

Chapter 2

Statutory Submission Requirements

Anna SHUM and Derek YUEN

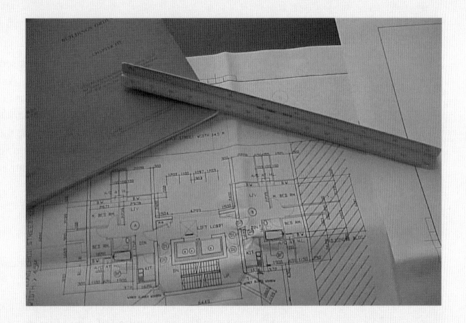

In Hong Kong, it is a statutory requirement that all proposed private building works must be submitted to the Building Authority for approval before their commencement on site. This chapter serves to provide a full picture of common practice in submitting drawings to the relevant government authorities for approval. It also explains the statutory submission requirements for new building works and the statutory procedures to be followed upon completion of building works. A conclusion is drawn on the latest development in submission requirements.

2

Statutory Submission Requirements

1 Introduction

In Hong Kong, it is a statutory requirement that all proposed plans for private building works must be submitted to the Building Authority for approval before their commencement on site. Building plans are prepared for submission after the client approves the design scheme presented by the designer. Government approval is required to ensure that the proposed building works will be carried out according to the minimum standards of public health and safety as laid down in the relevant legislation.

2 Major Legal Documents

Below are some major legal documents that are applicable to the development of a site (Figure 2.1):

a) Buildings Ordinance Cap. 123 governs the minimum standard of public health and safety required of a building development.
b) Outline Zoning Plan under the Town Planning Ordinance Cap.131 governs land use zoning.
c) Lease (non-statutory) governs the conditions of a lease agreement covering a piece of land.
d) Airport Height Restriction Map limits the maximum building height.

There are other ordinances that also affect the development of a site depending on the nature of the proposed development. For instance,

a) Education Ordinance Cap. 279 for schools;

b) Places of Public Entertainment Ordinance Cap. 172 for cinemas;

c) MTR Ordinance Cap. 276 for development near the MTR railway line; and

d) Environmental Assessment Ordinance Cap. 499 for environmentally sensitive development.[1]

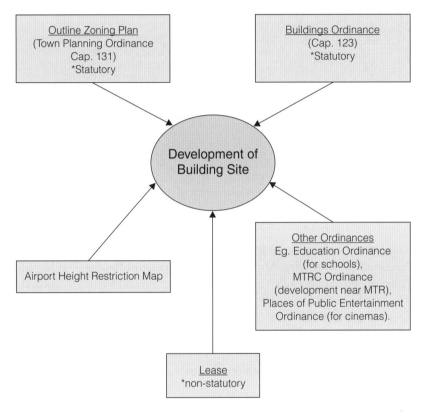

Figure 2.1   *Major considerations in the development of a building site*

For example, for the re-development of a site in Kowloon City, the architectural designer should check at least two major statutory documents affecting the development potential of a site, namely the Buildings Ordinance and the Outline Zoning Plan (with its Schedule of Notes), and the non-statutory document, namely the Lease. If, for example, the maximum allowable plot ratio under the Buildings Ordinance is 15 while that under the Outline Zoning Plan is 8 only, then the more stringent requirement should be adopted. The more stringent requirement takes density and facilities for infrastructure into more detailed consideration.

The airport height restriction has since relaxed after the airport was relocated. Since the buildings in Kowloon City can be built much taller

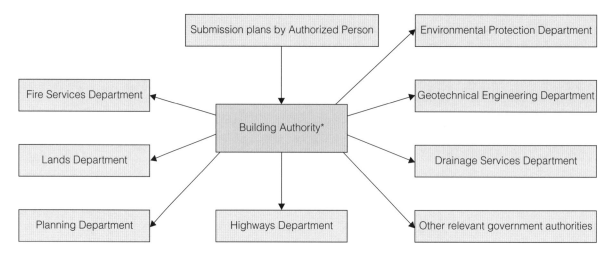

*Building Authority will distribute the submitted plans to various authorities and comments centrally collected back.

Figure 2.2   Central processing of submission plans

and denser, the burden on the infrastructure facilities of the area will be increased. The Building Authority will require an Environmental Impact Assessment Report to be submitted before approval to build is granted. The report should demonstrate that no adverse effect would happen to the nearby environment.

Furthermore, if the site is near an MTR tunnel, the MTRC Ordinance, which specifies additional safety and monitoring measures during construction, should be taken into account.

3   Central Processing System

Submission of building plans for government departments' approval follows a **central processing system**. The plans are submitted to the Building Authority who will then distribute the plans to the relevant government authorities for approval and/or comment.[2] The Building Authority itself will check the submitted plans for compliance with the Buildings Ordinance. Others, such as the compliance of Outline Zoning Plan, lease, fire services provisions etc, are distributed to the Planning Department, Lands Department, Fire Services Department respectively for approval/comment. The Building Authority will only disapprove plans if they have the power to do so as stipulated under the Buildings Ordinance (Figure 2.2).

The Central Processing System has helped to provide a more efficient communication means. The Building Authority will collect comments from all relevant departments and incorporate them into a single letter to the applicant. Very often, the comments so collected are attached to the Building Authority's letter of approval (or disapproval). This has often created confusion as some applicants may misinterpret this letter of approval as approval from all government departments, when it only means approval by the Building Authority for complying with the Buildings Ordinance. Other departments may disagree with the proposed building plans. An example is the proposed building plans submitted for an elderly centre. Despite the Social Welfare Department's lengthy written comment disagreeing with the proposed building plans, the Building Authority may still issue a letter of approval as long as the building plans comply with the fundamental issues of the Buildings Ordinance.

4   Statutory Actors and Their Duties

The Building Authority is empowered by the Buildings Ordinance[3] to approve or disapprove submission plans. These building plans for submission purposes are prepared by Authorized Persons. **Authorized Persons** (APs) are "statutory actors" who perform the duties and functions as stipulated under the Buildings Ordinance.[4] APs are the co-ordinators of building works.[5] Other major statutory actors include **Registered Structural Engineers** (RSEs), **Registered Contractors** (RCs) and **Registered Specialist Contractors** (RSCs). RSEs are responsible for the structural elements of building works[6], while RCs are responsible for carrying out the building works and SRCs for carrying out specialist works[7] such as demolition or asbestos removal (Figure 2.3).

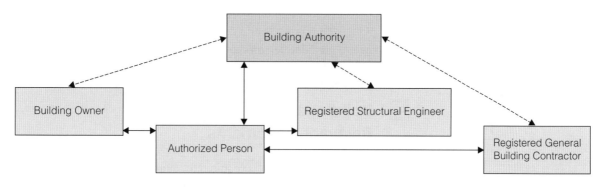

Figure 2.3   Communication among Building Authority, Authorized Person etc.

The duties of these statutory actors are:

a) to **supervise** the carrying out of the building works in accordance with the supervision plan (RCs to provide continuous supervision[8]);

b) to **notify** the Building Authority if there is any contravention of the Ordinance and its regulations; and

c) to **comply** generally with the Buildings Ordinance.[9]

The Building Authority keeps a register of these statutory actors (APs under the registered list of architects, engineers and surveyors). Applicants need to apply for inclusion and attend a separate interview conducted by the registration committee of the Building Authority who decides on the competence and experience of the applicants to perform the statutory duties stipulated under the Buildings Ordinance. For example, persons qualified as registered architects under the Hong Kong Architects Registration Ordinance do not automatically qualify as APs. Moreover, APs should be able to give professional judgment while exercising their statutory duties. The statutory duties relating to public safety (including building, structural and fire safety) is a matter of "life and death" for the occupants. An AP bears enormous professional liabilities for the entire life of the building built under his supervision.

Another example of the extent of APs' liabilities can be illustrated by the preparation of building plans. Building plans are prepared under the supervision of an AP. The plans are then submitted to the Building Authority for approval before any construction works can be carried out on site. The Building Authority will carry out curtailed check on fundamental issues only. The Building Authority has issued in its latest PNAP No. 272 (Jul 2002) a list of fundamental issues that will be checked by them. Major aspects include density, safety, health and environment, and other major issues in related regulations. The APs are wholly responsible for ensuring the Buildings Ordinance and its relevant regulations are complied with. The AP is liable for any contravention to the Buildings Ordinance found during or after a building's construction. This places a great burden on the APs.

5 Buildings Ordinance and Relevant Documents

Every **private** development is required to obtain the Building Authority's approval of the submitted plans and consent before commencement of building works on site.[10] Some buildings and building works are

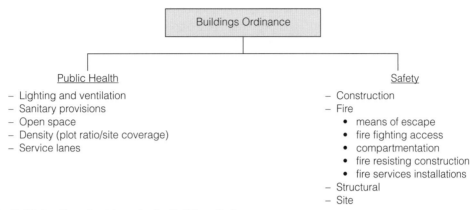

Figure 2.4   Public health and safety under the Buildings Ordinance

exempted from the provisions of the Buildings Ordinance. Examples are government or the Housing Authority's buildings, temporary places of public entertainment, building works not involving alteration to a building's structure and certain drainage works.[11] Although these buildings or building works do not have to follow the provisions of the Buildings Ordinance, they should not be constructed in contravention to the provisions of the Buildings Ordinance and its regulations.

The Buildings Ordinance lays down the minimum standard of public health and safety on a building site. When the designer is developing a detailed design scheme, he/she must take into account the various building ordinances and regulations. This ensures that a building will give its end users a minimum acceptable standard of safety and livability. **Public health** covers lighting and ventilation, provision of sanitary fitments and drainage, lane, open space, floor height, density, etc. **Safety** includes site safety during construction, structural safety and fire safety provisions (Figure 2.4).

The Buildings Ordinance identifies a number of **scheduled areas**[12] that require special measures to be carried out during construction due to their special nature, such as underground conditions. Examples are the Mid-Levels and the MTR protection zones.

The Buildings Ordinance includes a set of **building regulations** which elaborate on the various requirements of the Buildings Ordinance. For example:

a) Building (Administration) Regulations — the information required to be shown on submission plans, documents required to be submitted for approval, statutory periods of approval and consent, etc.

b) Building (Construction) Regulations — the prescribed mix of concrete, minimum loading capacity, etc.

c) Building (Demolition) Regulations — the prescribed method and procedure of demolishing an existing building.

d) Building (Planning) Regulations — the permitted maximum plot ratio and site coverage, minimum window areas for natural lighting and ventilation, fire safety provisions, minimum floor height, etc.[13]

The Ordinance and Regulations are passed as laws by the Legislative Council. The Building Authority has **Codes of Practice** that further elaborate on the technical requirements of the regulations. For example:

a) Code of Practice on the Provisions of Means of Escape in Case of Fire — the minimum number and widths of exit routes and doors, the maximum travel distances to escape from rooms to the exit routes, etc.

b) Code of Practice on Fire Resisting Construction — the fire-resisting materials permitted for construction, the minimum duration of fire-resistance required of different materials, etc.

c) Code of Practice on Overall Thermal Transfer Value — the thermal transmittance of building materials and façade design.

Practice Notes for Authorized Persons and Registered Structural Engineers (PNAPs) are issued by the Building Authority from time to time to clarify any 'grey' areas of the building laws and codes as well as the intention of any new laws or amendments to existing laws. PNAPs are a fast and effective way of communication to inform the APs about the latest interpretation of building laws (as they are issued by the Building Authority without going through any legislative process) (Figure 2.5).

Besides PNAPs which are issued by the Building Authority, various government departments also issue different practice notes to inform the Authorized Persons, and the general public, or clarify the departments' latest practices and interpretation of the building laws. These include the **FSD Circular Letters** issued by the Fire Services Department, **Lands Administration Office Practice Notes** issued by the Lands Department, and the **Practice Notes for Professional Persons** issued by the Planning Department. The Building Authority also issues **Practice Notes for Registered Contractors**. It is common practice that the Authorized Persons and other building professionals should keep themselves update with the latest practice notes and refer to these notes constantly while proceeding with the planning and construction of building projects, in order to avoid any mis-interpretation of the departments' practices or any improper procedures in gaining approval or consent from these departments.

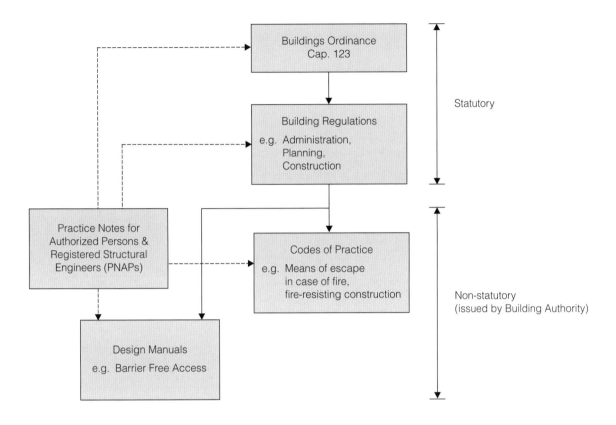

----▶ Building Authority's communication channel to clarify their interpretation of building laws and/or practice.

Figure 2.5   Relationship between Buildings Ordinance, Codes of Practice & Practice Notes

Most of the building laws are under review by the Building Authority to suit current and future building development. One such example is the encouragement of environmentally friendly provisions for buildings.

At the time of writing, the latest development in submission requirements has encouraged innovative ideas in the design and construction of, and submission procedures for, proposed building works. One example is the issue of **Joint Practice Notes**. For the first time, the Building Authority, Lands Department and the Planning Department have jointly issued practice notes for building professionals with the common aim of encouraging improvement and protection of the built environment. The Building Authority itself has set up an internal **Building Innovation Unit** (BIU) to review the existing legislation, to allow more flexibility when approving plans with innovative design and construction methods. The Building Authority is also considering a change of building codes from being prescriptive-based as at present to being performance-based. Moreover, most of the documents and forms issued by the Building

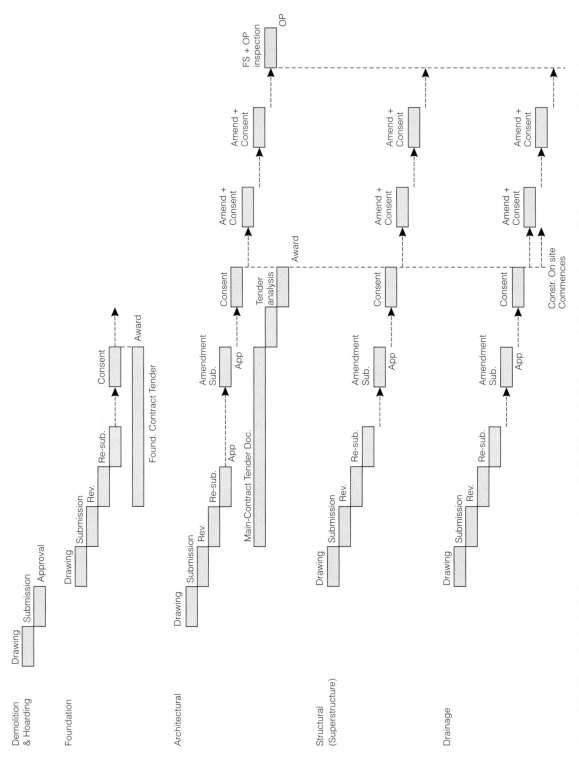

Figure 2.6   Example of submission programme for getting Building Authority's approval of plans and consent before commencement of site works

Authority are now accessible via the Internet. Some submissions in electronic format that are in compliance with the Electronic Transactions Ordinance have now been implemented. The Building Authority aims to set up an **electronic submission** system eventually to handle all types of electronic submissions.

The latest PNAP No. 272 explains the Building Authority's latest practice in providing a more effective building plan approval process. The Building Authority is prepared to hold pre-submission conference with various government departments to discuss, during the concept drawing stage, any unconventional design, structural principles/theories, or construction methods put forward by the building professionals before they proceed with the preparation of detailed design.

Laymen may find the various ordinances, regulations, codes of practice, practice notes, etc, as being too many and confusing. But the strength of the system is that the building laws are arranged systematically in a hierarchy. There is only one set of building laws to follow, although the interpretation of the building laws has created various disputes and appeals over the Building Authority's decisions, especially when the laws expressly give the Building Authority the power to exercise its discretionary rights.

6   Pre-commencement of Construction Works

Different plans are required to be submitted for **approval** at different stages of a building's development. Examples (in sequence) of these plans are (Figure 2.6):

a)   hoarding plan;
b)   demolition plan;
c)   excavation and lateral support plan;
d)   piling plan;
e)   building plan;
f)   framing plan; and
g)   drainage plan.

The Building Authority is empowered to approve or disapprove submitted plans. Reasons for disapproving plans include:

a)   the carrying out of works is in contravention of the provisions of the Ordinance;
b)   the carrying out of works is in contravention of the approved plans under the Town Planning Ordinance (for example, Outline Zoning Plan); and

c) the submitted plans require further particulars (documents) which have not been submitted.[14]

The Building Authority may also impose additional conditions while approving the plans.

Any person who contravenes the provisions of the Buildings Ordinance is guilty of an **offence** and liable to a fine and/or imprisonment, depending on the seriousness of the offence. The Buildings Ordinance provides a system of **appeal**[15] and review for applicants who are not satisfied with the Building Authority's decisions while exercising its discretion.

After the set of plans for each stage of development has been approved and before construction work commences, a separate consent is required for each separate stage of the works. This gives the Building Authority an opportunity to review plans that have been approved long ago (more than 2 years) and therefore may not comply with the latest building laws. In addition, the Building Authority has, since 1996, required the submission of a supervision plan before any consent is issued.[16] A supervision plan lays down the contractor's proposed construction site safety measures while carrying out construction works.

The Buildings Ordinance stipulates statutory periods for giving approval of plans and consent to commence building works. In general, the Building Authority will either approve or disapprove the newly submitted plans within 60 days from the date of submission. A consent (for example, drainage) can only be applied if the plans for such works have been approved. The consent will be issued within 28 days from the date of submission of the application. There are also fast-track processing procedures which take less time for obtaining approval of plans and consent to carry out building works. The approval of plans and consent can be granted concurrently, which typically takes 30 days only. An example is simple alteration works which do not involve the structure of the building and do not cause any violation of the Buildings Ordinance.

7 Commencement of Construction Works

It is not unusual to find changes in design while construction is underway. These subsequent changes in design would require amendments to the plans and re-submission for the Building Authority's approval, as well as consent for the approved amended plans, before the revised portion of works could be carried out. The new PNAP (No. 272) allows minor amendment works to be carried out on site without prior approval of these amendment plans. The submission of these amendment plans can be deferred until the application for an occupation permit/temporary occupation permit.

8   Completion of Construction Works

The Buildings Ordinance stipulates that no building can be occupied without obtaining an **Occupation Permit** (OP)[17] from the Building Authority. The OP enables the Building Authority to make sure that a building has been constructed according to all approved plans (as stated in the various consents to commence building works) and in compliance with all the provisions of the Buildings Ordinance before occupation is allowed. The Building Authority will issue the OP if the following conditions are met:

a) the building works have been carried out in accordance with the Ordinance;

b) the submission of a fire services certificate certifying the provisions of fire services equipment and installations;

c) the submission of a lift certificate, or if not, the liftway must be protected before the installation of the lift;

d) the submission of a water certificate related to the supply of water;

e) the submission of a performance review report if the site lies in the scheduled area;

f) the submission of the prescribed form for application of an OP;

g) the submission of record plans showing all approved works; and

h) the submission of a schedule of building materials and products, certifying their structural integrity and fire safety.[18]

Large sites may be developed in stages and therefore completed in stages. In this case **temporary occupation permits** (TOPs) can be obtained for phased completion. The above listed documents are submitted along with the application for TOPs. For example, it is common to occupy and use the shopping podium while the residential or commercial towers above are still under construction. TOPs are usually valid within a time limit. Applications for OPs or TOPs will be approved (or rejected) within a statutory period of 14 days from the date of submission.[19] Site inspections and testing by the Building Authority and the Fire Services Department will be arranged within the 14-day period.

Although there may be several TOPs, there is only one OP for each building site. The OP signifies the completion of building works in

Type	Relevant Document	Significance	Non-compliance
• Occupation Permit (O.P.)	Buildings Ordinance (Statutory)	AP to certify: • completion of building works • compliance with Buildings Ordinance • building is fit for occupation	Offence • fines & imprisonment
• Certificate of Compliance (C.C.)	Lease (non-statutory)	Lands Department to certify: • compliance with Lease conditions, including landscape, building, road, bridge, park, etc.	Breach of Lease • re-entry of site (worst case)

Figure 2.7   Occupation Permit and Certificate of Compliance

compliance with the Buildings Ordinance and its regulations and that the building is fit for occupation.[20] This document is totally different from a **Certificate of Compliance** (CC) in which the Lands Department (not Building Authority) certifies the compliance with Lease Conditions (not the provisions of the Buildings Ordinance). The OP is usually obtained earlier than a CC because the latter involves works other than building works, such as landscaping, bridges, roads, etc. (Figure 2.7)

9   Conclusion

Over the years, some practitioners have criticised the Buildings Ordinance and related legislation as being too rigid, out-dated and not reflecting current needs. Some people have argued that the inflexibility of the building laws has been an excuse for not producing innovative design. With the growing concern for environmental protection, the Building Authority has acted responsibly in recent years in encouraging innovative and energy efficient building design by, for example, allowing modifications to the existing stringent laws to cater for innovative design and construction.

Moreover, the building laws give the Building Authority the discretion to interpret the laws. These have previously created much misinterpretation and ambiguity, leaving many "grey areas" and loopholes in the laws. However, the Building Authority has been stepping up its efforts to convey their interpretation of the laws and clarify any "grey areas" by issuing Practice Notes for Authorized Persons and Registered Structural Engineers in a much more timely manner.

This is an exciting era in which major changes to the building codes and submission system are being made. It is important as a professional to keep pace with the latest development in the industry. Various professional institutes are very active in organising different continuing education programmes and professional development seminars to share views on the latest development in the industry with the professional institutes' members. This is one of the major channels through which professionals can keep up to date with developments in the building construction industry.

Notes

1. For other legislations and publications affecting the building industry, refer to PNAP No. 115 (revised Sep 2000).

2. Refer to PNAP No. 30 (revised Sep 2000) for different government departments and their functions in relation to the proposed building works.

3. Buildings Ordinance Section 16.

4. Buildings Ordinance Cap. 123 Section 3(2) stipulates that the Authorized Persons' register should contain a list of architects, a list of engineers, and a list of surveyors.

5. Buildings Ordinance Section 4(1)(a).

6. Buildings Ordinance Section 4(1)(b).

7. Buildings Ordinance Section 9(1) and 9(2).

8. Buildings Ordinance Section 9(5).

9. Buildings Ordinance Section 4(3).

10. Buildings Ordinance Setcion 14(1).

11. Buildings Ordinance Setcion 41.

12. Buildings Ordinance Schedule 5.

13. Refer to PNAP No. 115 for a full list of Ordinances, Regulations, Codes of Practice, Design Manuals, etc. that may affect the development of a building.

14. Buildings Ordinance Section 16(1).

15. Buildings Ordinance Section 44.

16. Buildings Ordinance Section 14A.

17. Buildings Ordinance Section 21.

18. PNAP No. 53 (revised in Jul 2002).

19. Buildings Ordinance Section 21(7).

20. Temporary occupation permit (TOP) therefore means that that portion of the building is in itself complying with the Buildings Ordinance, and is fit and safe for occupation.

References

1. Buildings Department. Practice Notes for Authorized Persons and Registered Structural Engineers (PNAPs). Last revised Jul 2002. Hong Kong Special Admininstration Region, from: http://www.info.gov.hk/bd

2. Buildings Ordinance Cap. 123 and Regulations. Last revised Jul 2002. Hong Kong Special Administration Region, from: http://www.justice.gov.hk

Chapter 3

Control on Unauthorized Building Works in Hong Kong

Anthony W. Y. LAI

Some existing building works are classified as unauthorized building works (UBWs) under the Buildings Ordinance and require statutory control. The Building Authority is empowered under the Buildings Ordinance to handle the UBWs problem. This chapter covers regulatory controls currently in force under the Buildings Ordinance and the government's policy on UBWs. An introduction to the relevant policies and controls, as well as an analysis of their impact on existing buildings, is followed by a review of the government's strategy for the removal of existing UBWs in Hong Kong. References are made to legal cases to illustrate the implementation of the regulatory principles.

PART I

CHAPTER

3

Control on Unauthorized Building Works in Hong Kong

1 Introduction

The pressure of continuous population growth and the influx of immigrants since the 1960s gave rise to a housing problem, which pushed up the demand for low cost housing. The scarcity of supply and lack of enforcement have finally resulted in the proliferation of **unauthorized building works (UBWs)** in Hong Kong, which include unauthorized rooftop structures and unauthorized structures attached to external walls and canopies. Figures 3.1 and 3.2 illustrate the common UBWs on the envelope of older buildings. These structures have degenerated over time and turned old urban areas into slums (Hong Kong Buildings and Lands Department 1991). In recent years, a series of incidents involving the collapse of alleged UBWs has prompted the government to look for ways to prevent the situation from worsening. A Task Force on Building Safety and Preventive Maintenance was consequently established under the Planning and Lands Bureau in 2000 to formulate a new policy for tackling UBWs.

This chapter will try to answer the following questions:

a) What kind of building works can be classified as unauthorized?
b) How is the Building Authority empowered under the Buildings Ordinance to deal with UBWs?
c) How successful is the current policy and what is the direction likely to be taken for tackling the UBWs problem in the future?

2 Definition of UBWs

According to the Buildings Ordinance, the term "building works" has a very wide meaning which includes all kinds of building construction, site

Metal cage attached on external wall

Illegal structure attached on external wall

Illegal structure may affect stability and increase site coverage and gross floor area of the main buildings

Figure 3.1   Proliferation of UBWs on envelope

Roof top structure creates fire escape problems

Figure 3.2   Erection of illegal structures (roof top structure)

formation, demolition, alteration and addition works, drainage works and building operation. "Building works" can be classified as unauthorized if they are carried out without prior approval of plans and consent for commencement from the Building Authority, unless they are exempted under Section 41 of the Buildings Ordinance. Under this regulation, exempted building works include:

a) buildings belonging to the Government;
b) buildings belonging to any person representing the People's Liberation Army;
c) buildings controlled and managed by the Housing Authority under the Housing Ordinance (Cap.283);
d) street or access road maintained by the Government;
e) drainage works, ground investigation or site formation works which does not affect the structure; and
f) building works of a non-structural nature or which do not contravene any regulation.

Buildings in categories (a) and (b) are usually managed by the Architectural Services Department. Item (f) gives rise to two questions:

i) What kind of building works are considered non-structural in nature?
ii) What kind of building works will contravene regulations?

To answer question (i), building works which have loading implications on existing buildings are considered "structural" and therefore requires the consent of the Building Authority. Non-structural partition walls are exempted under this definition. This interpretation however seems rather narrow. The rules as interpreted by the decision in the *Laguna City Case* (1995) support the definition of "structures" as being dependent on its size and type of connection (fixed or unfixed) to the roof, irrespective of its loading implications or the reason for the connection (Lai and Ho 2000). In particular, this case illustrates that the assembly of pre-set roof cabinets may be considered building works due to its size, construction and the method of attachment to the roof (Lai and Ho 2000). Accordingly, whether building works are considered structural not only depends on its structural impact on the existing building but also its construction.

To answer question (ii), it is necessary to consider whether the building works would be approved by the Building Authority if the plans were submitted for approval. According to the Practice Notes for Authorized Persons and Registered Structural Engineers (PNAP 272), the Building Authority will take four fundamental considerations into account in determining whether or not to approve the plans. These four considerations are:

a) density control within the development e.g. control on plot ratio and site coverage, etc;

b) the provision of adequate means of access for fire fighting and rescue; escape from fire; and fire resisting construction requirements;

c) the provision of reasonable amounts of light, air and open space;

d) fundamental issues under allied legislation, e.g. provision of adequate active fire protection facilities, Outline Zoning Plans requirements under the Town Planning Ordinance (Chapter 131), provision of adequate facilities for the disabled in accordance with the *Design Manual: Barrier Free Access 1997*, airport height restrictions and railway protection.

A review of the plans with these four fundamental considerations in mind would help to determine whether the proposed building works would contravene regulations and require the consent of the Building Authority. Commonly exempted works include extra lobby doors; entrance gates which do not obstruct the fire exit route; the removal of partition walls that are not used for fire compartmentalisation purposes; and sliding or collapsible gates.

3   Appeal Case Studies of Non-structural UBWs

Laguna City Case (1995)

Background

The appellant argued that the assembly of a cabinet for storage purpose with size 3,300 cm x 2,650 cm x 2,500 cm on a roof was not a "structure" and it did not create any loading problems on the roof, but merely a movable cabinet. The cabinet was fixed to the roof by means of thin metal strips bolted to the main structure to prevent it from being blown away during typhoon seasons.

Decision

The Tribunal held the view that the loading implications or the reason for fixing should be irrelevant for determining whether a roof cabinet is a "structure". Instead, the size and construction, such as the fixing or attachment methods to the roof should be the major considerations. The decision of this case therefore affirmed that the assembly of the roof cabinet did constitute building works.

The decision of this case strengthened the answer to question (i) above by a clear distinction between mere cabinets sitting on the roof or "structure" that involved building works due to its size, construction and the method of attachment to the roof and, not just its loading implication.

4 Enforcement Against UBWs under the Buildings Ordinance

The Building Authority is vested under the Buildings Ordinance with the discretionary power to enforce regulations against illegal structures in private sector to ensure that statutory requirements for safety and health are met after the private buildings have been certified for occupation. The discretion must be exercised in good faith and within the power delegated under the relevant legal principles, otherwise it will be regarded as *ultra vires* (i.e. beyond legal capacity) and be subject to judicial review.

5 Normal Demolition

Once the presence of UBWs has been identified, the Building Authority may take a series of actions to effect the removal of UBWs.

a) The Building Authority issues an advisory letter to request the cooperation of the building owners to remove the illegal works usually within an allowed period of 6 to 8 weeks. This advisory letter has no statutory effect.

b) If the above request is not complied with, a removal order under Section 24 of the Buildings Ordinance will be issued to require the demolition, removal and alteration of any building works which have been or are being carried out without approval and consent of the Building Authority and in contravention of the provisions of the Buildings Ordinance.

c) Owners will usually have two months to effect the removal of the UBWs. For non-compliance with this order, the alleged person shall be liable on conviction under Section 40 of the Buildings Ordinance to a fine of $50,000 and to imprisonment for 1 year. A further fine of $5,000 per day shall also be imposed for persistent non-compliance with the order.

d) If the order is not complied with, a government contractor will also be called upon to complete the works and the incurred cost, including a supervision charge, will be recovered from the owners.

e) Under Sections 44 and 47 of the Buildings Ordinance, owners have the right to lodge an appeal within 21 days of the order with the Appeal Tribunal and the enforcement of compliance with the order will cease until the appeal has been heard and a ruling made.

Due to this appeal mechanism, the process for the removal of UBWs will consequently be lengthened. There are many cases which challenge the discretion exercised in issuing the removal orders and whether they have been issued fairly and with due regard to all factors covered under Section 24. The *Discovery Bay Case* (1995) concerned the power of discretion conferred on the Building Authority to take action against illegal works under this section. The word "significant" as used in the 1988 policy, which classifies the removal of certain UBWs as high priority because "they pose an immediate danger" or "are very new" (Lai and Ho 2000). Both cases are considered the key points by the Building Authority. The decision on the *Discovery Bay Case* (1995) was followed by the *Marina Cove Case* (1996), where "significant and new" was the criterion (Lai and Ho 2000), which was also consistent with the priority for action set out in the 1988 policy (Buildings and Lands Department 1988). Due to limited resources, the 1988 policy is still in force to achieve the dual purpose of protecting public safety and containing the problem of UBW. This policy prioritizes UBW into two groups (Buildings and Lands Department 1991):

a) Immediate Enforcement: all newly constructed UBWs or those under construction and all UBWs posing a hazard to life or property are deemed "high priority" action for removal.

b) Prioritized Enforcement: all UBWs not categorized in the Immediate Enforcement group on which enforcement action will be deferred and taken in sequence.

With a wish to cultivate a positive attitude among owners toward voluntary removal of their unauthorized structures, the Building Authority will usually issue advisory letters as a first step to request the owners to play a co-operative role to remove identified UBWs. However the success rate seems very low as seen from the data published by the Building Authority. The number of advisory letters issued on UBWs were 10,486 and 30,016 in 1999 and 2000 respectively, but only 2,869 cases in 1999 and 1,059 cases in 2000 were resolved by the owners' compliance. This amounts to only 27% and 3.5% in 1999 and 2000 respectively (Building Department 2001b). Therefore a proposal has been made to upgrade the advisory letters to statutory warning notices to be registrable against the title (Planning and Lands Bureau 2001c).

6 Appeal Case Studies of Removal Orders

1. Discovery Bay Case (1995)

Background

Unauthorized works were first identified by the Building Authority on 30 April 1993, but the removal order was not issued until 6 May 1994. The appellant challenged the late issue of the removal order as unfair, since it was issued more than 6 months from the discovery of the offence which was considered to have lapsed the time limit for prosecution.

Decision

The tribunal dismissed the appeal on the following grounds:

a) Section 24 of the Buildings Ordinance conferred the power of discretion on the Building Authority as to whether or not to issue an order.

b) The works which were considered "significant" qualified the word "works" rather than the word "new". Therefore it was also fair and reasonable that if a UBW was considered not significant, irrespective of whether they were old or new, the Building Authority should have discretion whether to take enforcement action.

2. Marina Cove Case (1996)

Background

The argument in this case concerned the priority for the removal of the UBWs in question.

Decision

Justification was required from the Building Authority on whether the UBWs constituted "an imminently dangerous situation". As this could not be affirmed in this case, the Tribunal's decision would be based on whether the UBWs were "significant and new" at the material times. The Building Authority's order was set aside because the demolition of the UBWs could not be justified with reference to the policy of priority. One of the grounds of the Tribunal's decision was that the Building Authority failed to ascertain the UBWs as a new structure based on the visual site inspection.

The ruling in the *Marina Cove Case* (1996) was consistent with that laid down in the *Discovery Bay Case* (1995), which affirmed that the policy of priorities could be justified if either of the following criteria was met.

a) The structural stability of UBWs was in imminently dangerous situation.

b) The UBWs were considered 'significant and new' at the time of inspection.

7 Priority Demolition

With a wish to speed up the removal process, Section 24B of the Buildings Ordinance empowers the Building Authority to apply to the District Court for an order for the demolition or alteration of UBWs. No appeal mechanism is provided under this section since the owner can apply to the District Court for a hearing prior to the issue of the demolition order. There are four criteria for the issue of an order:

a) the UBWs constitute an immediate danger to life or property;

b) the UBWs are erected for commercial purposes;

c) the UBWs are situated in the common part of a building and are seriously detrimental to the amenities of the neighbourhood; and

d) the UBWs constitute a public nuisance.

Failure to obtain a demolition order does not preclude the issue of a removal order under Section 24.

8 Unauthorized Works-in-progress

When the illegal building works are in progress, which is always treated as a high priority category for immediate enforcement action, the Building Authority is empowered to issue a cease works order under Section 23 of the Buildings Ordinance to stop the works as quickly and as safely as possible. Usually the Building Authority will issue a removal order under Section 24 after the cease works order, demanding the completed parts of the unauthorized works be demolished or rectified to the satisfaction of the Building Authority. It is not uncommon for the Authorized Person to be approached by the owner to submit plans for the completed building works for approval. However, there are a number of appeal cases e.g. *Sylie Road Case* (1995) *and Filipino Club Case* (1995), which affirm that the

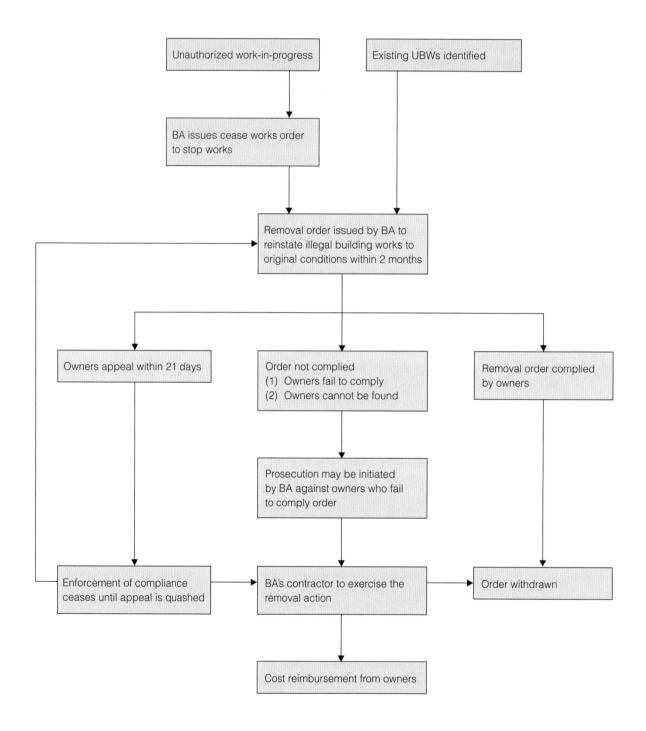

Figure 3.3   Enforcement of control on UBWs under Buildings Ordinance

Building Authority has no power to give retrospective approval or consent for building works which have already been completed. The decisions on these cases are consistent with the Practice Note for Authorized Persons and Registered Structural Engineers (PNAP 125), which advises that the Building Authority will only entertain prior approval of plans and consent for commencement. What the Authorized Person should do under this situation is to help the owner identify the completed sections of illegal works to be removed or rectified to the satisfaction of the Building Authority prior to the submission of plans for approval.

The flow chart 1 in Figure 3.3 illustrates the enforcement actions vested on the Building Authority for dealing with UBWs and three different scenarios for the owners to response to a removal order issued by the Building Authority.

9   Tribunal Appeal Case Study of Retrospective Approval for Completed Building Works

Sylie Road Case (1995)

Background

This case involved an appeal against the Building Authority's rejection of an application for approval for plans relating to an existing unauthorized temporary structure.

Decision

The Tribunal concluded that the Building Authority was in a right position in refusing to process approval of plans for building works already commenced or completed.

10   Court Case Study of Retrospective Approval for Completed Building Works

Filipino Club vs Building Appeal Tribunal and Anor (1995)

Background

A judicial review was sought regarding a demolition order requiring the removal of an illegal car-parking platform which had been added to the club premises. The remedial work was required to be under the supervision of an authorized person (AP). The AP submitted plans for approval on normalizing the completed illegal structure, but the Building Authority refused to provide retrospective approval on buildings works that had already been completed.

Decision

The judicial review was dismissed because application for approval should be made prior to the construction of the illegal structure in question. Therefore the Building Authority was entitled to reject this retrospective approval.

The decisions of these two cases purportedly indicated that the AP is obliged to supervise the reinstatement of the completed illegal works to original condition prior to the submission of plans for approval.

11   Closure Order

The Building Authority could also, under Section 27 of the Buildings Ordinance, apply for a closure order from the District Court if the UBWs:

   a)   render a building dangerous or potentially dangerous; or
   b)   require the closure of the building for carrying out remedial works without causing danger to the occupants or the public.

The first category usually refers to UBWs which pose a serious threat to the safety of the occupants, neighbours or the public because the building has structural problems or is in a dilapidated state. UBWs which obstruct fire escape routes fall under the second category. UBWs which may commonly lead the Building Authority to apply for a closure order are rooftop structures or cockloft door openings in the fire exit staircase of single-staircase buildings.

12   Measures to Tackle UBWs

Enforcement: Large scale clearances

Despite various measures to tackle the problem, UBWs on the external walls and roofs of older buildings continue to exist, worsening the deterioration of buildings and posing a threat to the public. In 2000, the

Building Authority estimated there were about 800,000 UBWs in Hong Kong, including illegal rooftop structures, the type of UBWs which most often gives rise to long-term social problems (Planning and Lands Bureau 2001c).

To prevent an increase in the number of UBWs, the Building Authority has adopted an immediate removal policy for newly erected UBWs. In parallel with this policy, the Building Authority has recently implemented two large scale operations:

a) "Blitz" operations will kick off from time to time with an aim to identify individual buildings of 20 to 40 years old for large-scale clearance of UBWs, including those on podiums, rooftop or in yards and back lanes. These targeted buildings, usually located along heavy traffic routes, may lead to serious consequences in case of any disrepair of the UBWs' building elements. This approach has proved to be an effective deterrent for individual flat owners and has considerably improved owners' compliance. These operations aim at removing 150,000 to 300,000 UBWs in five to seven years (Planning and Lands Bureau 2001c).

b) Coordinated Maintenance of Buildings Scheme (CMBS) is a more comprehensive proactive approach to tackle building maintenance and UBWs in one operation. In order to strengthen support for building owners, the CMBS is a "one-stop" service approach. It provides assistance in contract procurement and tendering services for the engagement of appropriate building professionals and contractors for building maintenance and removal of UBWs. This scheme involves the collaboration of five departments, namely the Home Affairs Department, Fire Services Department, Electrical and Mechanical Services Department, Food and Environmental Hygiene Department and Water Supplies Department (Buildings Department 2000). The Buildings Department (BD) acts as a Building Coordinator under this scheme, similar to the role of an Authorized Person (AP) in a private alteration and addition project, to liaise with different professional bodies and provide technical advice and assistance to encourage participation from building owners in determining how to deal with UBWs on building envelopes and maintain a building properly.

These two large-scale operation approaches are believed to be effective in curbing the increase in the number of UBWs and are supported by the Task Force on Building Safety and Preventive Maintenance (Planning and Lands Bureau 2001a). According to the data published by the Buildings Department, from 1999–2001, the total number of removal orders issued by the Building Authority was 14,970 in 1999, 9,469 in 2000, and 12,854 in 2001, and the number of owners' compliance with removal orders was 3,684 in 1999, 9,139 in 2000 and 11,135 in 2001 (Buildings Department 2001b). The success rate was significantly increased from 24.6% in 1999 to 96.5% in 2000 and 86.6% in 2001. These figures reflect that the large-scale operations have achieved the deterrent effect for owners to comply with the statutory orders issued by the Building Authority.

13 Enforcement: Roof-top Structures

Illegal rooftop structures in single-staircase buildings are considered potential "fire-traps" (Planning and Lands Bureau 2001a) the erection of which violate the requirements of the Code of Practice for the Provision of Means of Escape In Case Of Fire. Buildings with fewer than six storeys are permitted to have only one staircase while buildings with more than six storeys are required to have two staircases for fire exit purposes under the Code of Practice. Thus, when unauthorized rooftop structures increase the number of storeys to more than six, they increase the fire risk of those buildings and are classified as UBWs of the high priority group under the policy of the Buildings Department. The procedures for removal of unauthorized roof-top structures are similar to those described above with the exception that upon the expiry of Section 24 removal orders, the Building Authority will resort to the application for a Closure Order with the District Court rather than directly proceeding to removal through government contractor action. This is considered necessary to avoid confrontation with the residents of the roof-top structures and ensure the whole or part of the building will be closed for demolition without danger to the occupiers and to the public. Since these illegal rooftop structures are usually occupied by low income groups such as immigrants from the Mainland, collaboration with other government departments i.e. the Housing Department, the District Office, the Police and the Social Welfare Department, is necessary prior to their removal so that the affected occupants may be rehoused and unnecessary confrontations avoided (Planning and Lands Bureau 2001a). The HKSAR Government is also considering a proposal to prohibit transactions on illegal rooftop properties through an amendment to the Conveyancing and Property Ordinance and related legislation (Planning and Lands Bureau 2001c).

14   "Encumbrance" on Property

Once the orders mentioned above have been issued by the Building Authority, they will be registered in the Land Registry as an "encumbrance" on the property. "Encumbrance", as defined in the case of *Lam Mee Hing vs Chiang Shu Yin* (1995), is a claim to the property or charge which could be imposed upon the property. The encumbrance will remain until the orders are fully complied with to the satisfaction of the Building Authority. If a purchaser concludes a transaction for a property with an encumbrance on it, he/she will then be responsible for the encumbrance. The presence of UBWs will constitute a defect in the title of the property, as shown in several court cases, e.g. *Giant River Ltd vs Asie Marketing Ltd (1990) 1 HKLR 297 and Ip Cho Sau & Another vs Leung Kai Cheong & Others HCMP No 1288, 27 January 2000*. The owners' attention is drawn to the presence of UBWs which in some cases may affect the acquisition of the title (*Ip Cho Sau & Another [2000]*). Moreover, the HKSAR Government is now considering requiring owners to make a self-declaration prior to the transfer of title that there is no UBWs on the external walls of their properties (Planning and Lands Bureau 2001c).

15   Court Case Studies of Encumbrance

1.   Lam Mee Hing vs Chiang Shu Yin (1995)

Background

One of the arguments in this case was whether a building order issued by the Building Authority, which required remedial works to be carried out, could result in encumbrances against the title of the affected flat leading to a purchase agreement being rescinded.

Decision

The Court was satisfied that a building order which required remedial works could incur expenses to the owners of the flat and such expenses could become a charge to the property. This charge arising from the building order could be discharged if the vendor, the defendant, undertook the responsibility for the apportioned costs incurred from the remedial works.

2. Giant River Ltd. vs Asia Marketing Ltd. (1990)

Background

This case was related to the repudiation of a property transaction agreement because of an allegation that the defendant could not adduce good title to the property due to the extensive unauthorized structures on the property.

Decision

The existence of unauthorized structures required the government to take enforcement action and constituted a defect in title.

3. Ip Cho Sau and Another vs Leung Kai Cheong and Others (2000)

Background

This case concerned the intention of the purchaser to repudiate a sale and purchase agreement on the ground that good title could not be adduced due to the existence of unauthorized structures on the property.

Decision

The decision applied the ruling in Giant River Ltd v Asie Marketing Ltd (1990)(case no. 2, above) *that the presence of unauthorized structures together with a risk of enforcement action by the Government constituted a defect in title and an encumbrance was therefore created on the property.* However, the vendor was allowed to make the defect adduce good title by the time of completion of the agreement and the purchaser should not be in a haste to terminate the agreement immediately.

The above three Court cases purportedly indicate that the presence of unauthorized structures do constitute a defect on the title of the property. Any building order issued for its removal will be registered in the Land Registry and will create an encumbrance as a charge to the property.

16   Fast-track Approval of Minor Building Works

In addition to the above proactive approaches towards the removal of UBWs, amending the approval and consent process for building works is also an indirect way of encouraging owners to seek approval for minor alterations and/or additional (A&A) works. Approval of building plans normally takes a minimum period of 60 days. Including 28 days for obtaining consent for the commencement of building works, the approval and consent period takes approximately three months. For minor A&A works, the approval and consent period may be longer than the construction period, which may be as short as one month. Such A&A works are considered minor so owners often think they can proceed even without going through the approval process. After the introduction of the fast-track processing for minor, non-structural A&A works in 1997, the approval and consent can now be processed at the same time and completed within 30 days. For some minor works, such as simple amenity features, e.g. lightweight canopies and drying racks, the government is under consideration to allow private certification of their safety by professionals or contractors and this provision may also be incorporated in a future amendment on the Buildings Ordinance (Planning and Lands Bureau 2001e).

17   Education on Owners' Liability

The government always stresses that it is the owners' responsibility to maintain a safe environment free of UBWs. Promoting positive attitudes toward owners' responsibility should not just focus on ensuring the soundness of one's premises, but also on fulfilling one's responsibility to the community. To foster a stronger sense of owners' liability, education is of prime importance. In this respect, the government has adopted a multimedia channel to educate the community. The Home Affairs Department, in collaboration with tertiary institutions, also organizes a lot of training courses for owners' corporations to strengthen their sense of building safety, respond to enquiries and resolve disputes related to the removal of UBWs (Planning and Lands Bureau 2001d).

18   Revised Policy and Implementation Plan

The recent resolution of the Task Force on Building Safety and Preventive Maintenance recommends the 1988 enforcement policy against UBWs be revised to broaden the range of high priority UBWs for immediate enforcement action. Following this revised policy, the implementation plan as formulated by the Building Authority mainly focuses on intensifying enforcement actions on newly constructed UBWs including illegal alteration to newly designed environmental features in accordance with the Joint Practice Note 1 on "Green and Innovation Buildings", substantial UBW items constituting imminent danger to life or property and targeted buildings with extensive UBWs for large-scale clearance operations, e.g. blitz clearance of UBWs and the Co-ordinated Maintenance of Buildings Scheme (Planning and Lands Bureau 2001c). In order to make the best use of available resources to accelerate the removal of UBWs and to enhance cost-effectiveness, blitz operations to clear UBWs on external walls will be carried out through outsourcing of contracts to employ private building consultants to carry out investigation work and conduct compliance inspection after the issue of removal orders (Buildings Department 2001a).

19 Conclusion

The UBWs problem has persisted despite the effort made by the Building Authority. There is no single solution that will instantly resolve the problem. Instead, a collective effort by various government departments and building owners is required (Planning and Lands Bureau 2001b). Since UBWs is a socio-economic problem rather than a technical or design problem, enforcement under the Buildings Ordinance, though effective, will not eliminate it. The HKSAR Government is aware of this and has adopted a multi-pronged and partnership approach towards the problem (Planning and Lands Bureau 2001b). The launch of the Coordinated Maintenance of Buildings Scheme, which involves five departments as stated above, is a very good example.

The multi-prong approach towards the UBWs problem can be conclusively viewed to include legislation, proactive removal action and education. To strengthen the deterrent through legislative approach, two ordinances may be amended in the near future. Changes can be made to the Buildings Ordinance to allow private certification of minor works, upgrade the advisory letter to a warning notice with statutory effect which can be registered against the title of the property, and increase the penalties for non-compliance of statutory orders. The other is the Conveyancing and Property Ordinance. Changes may include a ban on conveyancing and letting of illegal rooftop structures.

The adoption of immediate removal policy for newly erected UBWs and the implementation of a series of blitz operations on selected target buildings for large clearance of UBWs on external envelope have achieved the deterrent effect and more effectively enforced owners' compliance on the removal orders.

Prevention is always better than cure. Effective building management through the implementation of the Building Management (Amendment) Ordinance 2000 can help curb the problem of UBWs. The Home Affairs Department can also provide professional assistance in persuading owners to remove UBWs or refrain from erecting new UBWs (Planning and Lands Bureau 2001b).

In view of a number of appeal cases challenging the fairness on exercising the discretion on priority treatment on UBWs, it is suggested that the transparency of the policy of prioritizing UBWs for action should be increased so as to enhance public awareness on high priority group UBWs.

Apart from these measures, education through seminars and campaigns are also recognized as a long-term investment in promoting public awareness of the UBWs problem.

All the above measures have reflected the government's strong commitment to stop the proliferation of UBWs. Although it is unrealistic to expect total elimination of all UBWs, checking their growth and removing imminent or potential danger arising from UBWs remain a realistic long-term objective. However, its fulfilment requires continuous collective action and stringent enforcement of the Buildings Ordinance. All the proposed strategies mentioned will require an increase in manpower and financial resources, which may be considered an investment in public safety. It is possible to create a safe built environment with an attractive cityscape through the government's commitment to this goal.

References

1. Building Appeal Cases. Lands Tribunal of Hong Kong.

2. Buildings and Lands Department. 30 March 1988. The control of unauthorized building works under the Buildings Ordinance, Cap. 123. Press release. Hong Kong: Hong Kong Government.

3. Buildings Department. 1996. Code of Practice for the Provision of Means of Escape in case of Fire. Hong Kong Government.

4. _____. 1996 & 2002. Practice Note for Authorized Persons and Registered Structural Engineers. 272 (2002) & 125 (1996). Hong Kong Special Administrative Region.

5. Buildings Department. 2000. An Introduction to the Coordinated Maintenance of Buildings Scheme. Electronic Brochure.

6. Buildings Department. 2001a. Contracts awarded to clear unauthorized building works. Press release.

7. Buildings Department. 2001b. *Monthly Digest*. Hong Kong Special Administrative Region.

8. Buildings Department, Lands Department and Planning Department. 2001. Joint Practice Note No. 1. "Green and Innovation Buildings." Hong Kong Special Administrative Region.

9. Buildings Ordinance Office. 1991. *Building Control in Hong Kong*. Hong Kong Buildings and Lands Department, Hong Kong Government.

10. High Court — Miscellaneous Proceedings. High Court of the Hong Kong Special Administrative Region.

11. Hong Kong Government. 1996. Buildings Ordinance, Chapter 123.

12. Lai, Lawrence, W. C. and Ho, Daniel, C. W., ed. 2000. *Planning Buildings for a High-Rise Environment in Hong Kong — A Review of Building Appeal Decisions*. Hong Kong: Hong Kong University Press.

13. Planning and Lands Bureau. January 2001a. Legislative Council, Panel On Planning, Lands and Works, "Building Safety And Timely Maintenance — To tackle unauthorized building works." Discussion Paper.

14. _____. 2001b. Maintaining buildings by multi-prong approach and partnership. Press release.

15. _____. 2001c. Task Force on Buildings Safety And Preventive Maintenance. Implementation Plan.

16. _____. 2001d. Government pledges action with assistance for building safety. Press release.

17. _____. 2001e. Government determined to tackle unauthorized building works. Press release.

18. Walden, Philip. 1995. "Structures, legal and illegal (or why things somethings get taken down by the Buildings Ordinance Office)." In *Building Journal Hongkong China*. November 1995.

Building Appeal Tribunal Cases

1. *Discovery Bay*. 1995. Building Appeal Case No.: 38/94.

2. *Laguna City*. 1995. Building Appeal Case No.: 08/95.

3. *Marina Cove*. 1996. Building Appeal Case Name: House No. F10, Marina Cove Stage II, Sai Kung, New Territories.

4. *Sylie Road*. 1995. Building Appeal Case No.: 35/94.

[Detailed explanation of these cases is given in Lai and Ho 2000.]

Court Cases

1. *Filipino Club v Building Appeal Tribunal & Anor* (1995) High Court — Miscellaneous Proceedings. No. 977 of 1995.

2. *Giant River Ltd. v Asia Marketing Ltd.* (1990) High Court — Miscellaneous Proceedings. No. 2510 of 1987.

3. *Lam Mee Hing & Anor v Chiang Shu Yin* (1995) High Court — Miscellaneous Proceedings. No. 1866 of 1995.

4. *Ip Cho Sau & Another v Leung Kai Cheong & Others* (2000) High Court — Miscellaneous Proceedings. No. 1288 of 2000.

Part II
Design and Management

Chapter 4

From Finding Form

Joseph Francis WONG

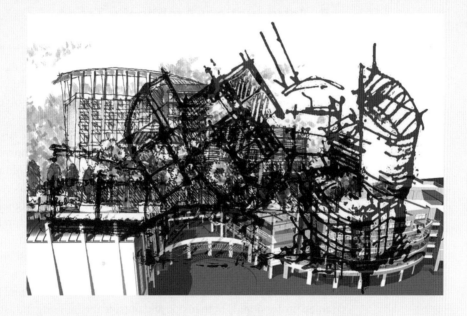

One of the skills architects possess that sets them apart from other trained professionals both inside and outside the building industry is space design — their ability to arrange cells of space of various forms and shapes into groupings that serve certain purposes, utilitarian or otherwise. However, architectural design is often perceived as an intuitive process — people see imagination and creativity as the main driving forces behind a successful design because of their initial attraction to the visual aspects of works of architecture. This chapter gives an introductory account of some of the activities that take place in the early stages of the building design process.

4

From Finding Form

1 Introduction: Architect's Trade

Without question, architectural design is decidedly a multidisciplinary activity. In various stages of a building's design, the designer will in varying degrees of detail tackle issues regarding structure, material, climate, function, building service, construction, etc., and solve problems arising thereof. Of course, at least nowadays, there are scarcely any architects who would call themselves experts in all of the above fields. Instead, they rely heavily on the advice of a team of consultants of a multitude of disciplines to complete the design of a building, not to mention the latter's increasing involvement in the process as construction on site commences. The likes of visionary Spanish designer Santiago Calatrava, who was trained both as an architect and a structural engineer, is a rare breed these days. That the architect is dependent on so many experts of other disciplines comes as no surprise as architects are often categorized as the practitioner of all trades but an expert in none.

One of the skills architects possess that distinguishes them from other trained professionals both inside and outside the building industry is space design, i.e. their ability to arrange cells of space of various forms and shapes into groupings that serve certain purposes, utilitarian or otherwise. Probably ten out of ten architects when asked what skills and knowledge set them apart from experts of other fields would claim it is their ability to create space, or their understanding of morphology, where morphology can be defined as the science of shapes. Although it remains a paradox that what an architect actually designs, and eventually become constructed, are the boundaries — walls, floors, windows, etc. — that define space rather than space itself. The spatial aspects of architecture constitute the focus of research as well as discourse of architectural knowledge. When given a design brief (a list of the client's requirements and desires), the architect can transform it into architectural ideas taking on various forms. But curiously, although the design follows directly from the design brief, there is no simple derivative relationship between the two. Such a simple relationship entails that the same set of design brief would provide the same design. On the contrary, giving two architects identical design briefs is always guaranteed to produce two completely different solutions. This is evident in design competitions where a full range of entirely different, even conflicting, solutions is derived from exactly the same information given in the competition guidelines.

Figure 4.1   The "Black Box" design process

Architectural design is often perceived as an intuitive process by the uninitiated (and the initiated as well some may argue), such as beginning students in design studios. They see imagination and creativity as the main driving forces behind a successful design because of their initial attraction to the visual aspects of works of architecture. They would approach design problems as some kind of black box operation where an abstract brief is fed into one end, and after some magical operation, a concrete architectural solution is spat out the other (Fig 4.1). Inevitably, this is unfortunately how students with no prior design experience tackle their studio work. Like the Zen calligrapher, they would stare at a blank piece of yellow tracing paper for hours with a 6B pencil in hand, waiting for the moment when inspiration strikes. Unfortunately, they often find the first line extremely hard to come by. If you ask them what are they doing, they would invariably reply: "I'm thinking about my design." To that, I would say, "you cannot see what you think." Admittedly, any design activity can be more related to the designer than the actual design problem. As Gunnar Birkerts puts it: "The creative act is a very personal event and a lonely one, almost like birth or death" (BIR ix). But is design really such a singular event that involves no reasoning? How do architects find form?

2   Inside-Out + Outside-In

Although space and form are the factors that most visibly distinguish one building from another, they are only two of many issues that an architect must consider when designing a building. Architecture is a multi-disciplinary field. The realization of a building project involves the input of skills and knowledge from a whole range of disciplines, from arts and humanities to science and technology. The architectural designer, though not expected to master ALL related disciplines, must have a thorough understanding in a number of them.

If you look at all the factors that may affect the design of a building, some of them are forces acting from the inside of the building (e.g. structure system, circulation, functions, etc.) while others are working on the outside (e.g. context, views, statutory requirements, etc.) (Fig 4.2).

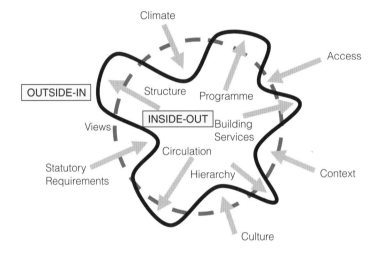

Figure 4.2   The inside-out and outside-in forces of architectural design

Looking back at the earlier assertion that architects will come up with different designs following the same set of design requirements points to the notion that the architects themselves are also adding something to the design, on top of the design brief. So when designing a building, architects do not distance themselves from the design and work entirely from the outside of the project. Instead, they combine some attributes coming from inside themselves with the requirements of the design brief, which originates from outside of the architects.

Philip Steadman refers to architectural design as selection of form and cites four factors affecting the architects' choice of form: aesthetic purpose, utilitarian requirements, technical possibility, and morphology, i.e. the general science of possible forms. Knowledge pertaining to the first three aspects is covered in courses independent of the design studio: aesthetics in courses covering history, the classical orders, proportions; utilitarian aspects in building type analysis, anthropometric studies, statutory controls; and technical problems in construction technology, structures, building services. Problems specific to morphology, however, are seldom given the same treatment as the other fields of knowledge in architecture. This further mystifies the architectural design process.

3 Design Process

How do architects begin the design of a building? Do they instinctively start dividing a form into pockets of space for different functions? Hardly, though some architects may very well do. Most do not. American architect Cesar Pelli, who is world-famous for his many high-rise creations such as the Petronas Towers in Kuala Lumpur, begins his projects not by designing or drawing. His approach to a project is not to try to immediately come up with any kind of formal solutions but to rely on various means of analysis to gain a better understanding of the characteristics of the project. This may take the form of site analyzes and visits, endless meetings with the client and users to discuss objectives, study of the physical context with the help of models and diagrams, discussions with relevant statutory bodies, etc. All these are aimed at obtaining more information on the design problem so that the architect begins to know which questions to ask rather than try to prematurely answer any questions. This is the formation stage of the design process.

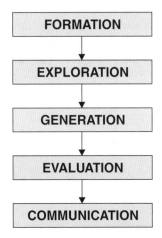

Figure 4.3   The different activities in the design process

We all know that architects are well-known for having their own peculiar ways of doing things. This is most evident in the mind-boggling diversity of their creations. Every architect has their own specific approach to solving a design problem or, for some of them, an array of strategies they use when facing different types of design problems.

Despite the huge variation in design approaches from architect to architect, there are a number of activities in the design process that are common to most of them: formation, exploration, generation, evaluation and communication (Fig 4.3).

The following summarizes some of the activities of this design process and what may take place when an architect is designing a building in Hong Kong:

A. Formation

After receiving the brief from the client, many constraints and criteria governing the problem is not known or at best loosely defined. At the inception of a design problem, when the design brief is initially set, there is not a definite concept of the final solution. The architect's first task is therefore to study the many implications of the given problem and identify the constraints limiting the number of possible solutions and the criteria against which these solutions will be judged.

One of the things the architect may want to find out at this stage is the development potential of the target site, i.e. what type of building can be built and the maximum floor area allowed. In Hong Kong for example, architects would check the development potential of a site as governed under three separate documents: the Building Ordinance, the Outline Zoning Plan (OZP) and the Government Leases, which control various aspects pertaining to the kind of buildings that can be built from the macroscopic to the microscopic level. These initial studies are normally followed by the first of many visits to the site for a preliminary site analysis.

B. Exploration

After forming the preliminary framework of the problem, the architect can then explore all the available sources of particular information and general information relating to the problem. Particular information is information specific to the project on hand only, e.g. the dimensions of the site, surrounding views, the history of the site, the requirements of the client for this particular building, existing buildings, etc. This type of information varies from project to project. On the other hand, general information is information that applies to all projects of a similar nature. These include the local building codes, recognized anthropometric standards, structural implications, etc. All the collected information is then analyzed.

Architects can search for land records at the Land Registry, which is set up to provide land registration and search services to facilitate property transactions in Hong Kong. All Government Leases as well as any extensions or modifications are on record and are open to the public. The 11 sections of the Hong Kong Planning Standard and Guidelines published by the Hong Kong Planning Department provide information on the government's criteria for determining the scale, location and site requirements for various land uses and facilities and guidelines for the design and planning of the built environment.

Apart from preliminary checking against the statutory and technical requirements for various aspects of the building design, one may also explore design ideas with the aid of "concept sketches". Everybody can come up with great ideas from time to time. However, great ideas do not automatically become great architectural designs. Innovative ideas must be further developed to examine their architectural implications and how they can be applied in an actual building design. This is where concept sketches can be of immense help. One of my former teachers at graduate school constantly reminded me: "You cannot see what you think!" So he was always encouraging us to sketch out our ideas, even if it was just an abstract idea.

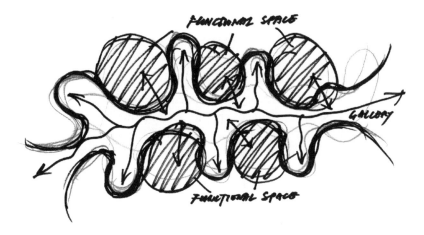

Figure 4.4   Concept sketch for an art school

For example, figure 4.4 above shows an early concept sketch for an idea for a competition for the Art School of the Hong Kong Arts Centre. The concept was that a gallery for an art school should be a place for exchange and integrated with the rest of the school's functional spaces to create interaction. The conventional type of gallery with a cellular layout would be too isolated from the rest of the school and only serves to separate the public from the creators. As you would see later on, although this sketch is nothing like a building plan yet, it nonetheless helped set an objective for the design development that followed. More importantly, in order to draw out your early ideas you need to put concepts into concrete images. This way, creating concept sketches is not only a means to record your ideas but also a means to develop your early abstract ideas into solid concepts — sketching is thus also visualizing.

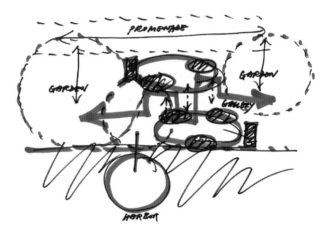

Figure 4.5   Bubble diagram for an art school design

C. Generation

With all the gathered and analyzed information, possible solutions can be generated. It is however impossible to generate the "perfect solution" directly from the information present thus far. Many possible solutions will be generated and developed before becoming design proposals to be considered. One common way a lot of architects use in starting a design is the generation of "bubble diagrams". Bubble diagrams are literally diagrams made up of bubbles (circles or whatever shape one may wish to use), which represent the various spaces in a design. These are normally joined together with simple lines representing relationship between these spaces.

A bubble diagram for the design for the art school competition is shown in Figure 4.5. This kind of bubble diagrams is a helpful tool in the early stages of design for working out fundamental relationships between the various parts of a building. Instead of starting off by drawing precise walls directly to demarcate separate rooms of a building, such spaces are abstracted and are represented by various shapes and line types to make the relational aspect of the design clearer to the architect. If all architectural elements — walls, doors, windows, floors — are drawn out, there would simply be too much information to digest and would take up too much time to sketch. Such information can be added on progressively as the concept of the design becomes clearer and clearer over a number of rounds of bubble diagrams.

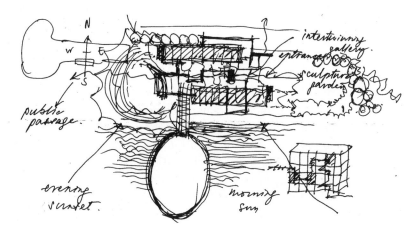

Figure 4.6   A more developed design diagram

Figure 4.6 shows a diagram for the Art School after a number of rounds of transforming the previous bubble diagram. You can see that the principal components and their basic relationships from the diagram in Figure 4.5 remain largely intact. Each component, however, are being depicted in more and more detail and their preliminary forms begin to emerge. It is a good practice to keep all these earlier sketches and diagrams. They not only represent the thought process behind the development of the design thus far but more importantly this set of diagrams becomes part of the criteria that the design should be evaluated against as the design process progresses.

D. Evaluation

There can be two types of evaluation in a design process: the generated design proposals can be evaluated against a pre-determined set of criteria or the criteria themselves will be evaluated when more information on the design is obtained through the development of the design itself. Therefore, not only the design proposal is continuously being transformed by the architect as more and more factors are taken into consideration, the designer's understanding of the problem also changes as various aspects relating to the problem are revealed as the design process progresses. In this sense, the formation of the design problem in stage A is not definitive: designing is the uncovering process that helps form the definition of the design problem and its solution.

Normally one would associate evaluation with some kind of conclusion of a stage at which point the results will be judged against requirements or objectives of the design problem. This kind of evaluation takes place often in architectural design during desk-crits and reviews in school and during meetings and presentations in the office. Such explicit evaluations almost always involve external parties in the evaluation process — tutors, guest jurors, senior members of the office, clients, etc. — and presentation materials (as opposed to design materials such as study model, sketches and diagrams) are specially prepared for these occasions. In explicit evaluations, the designer receives and responds to comments and questions from the evaluators and, if you are a student, hopes to pick up valuable points to improve on your design skills and, if you are a professional, improve on your design.

There is however another kind of evaluation — implicit evaluation — which is often overlooked even though we do a lot more of it as designers. As mentioned in the beginning of this chapter architectural design is by nature a multidisciplinary field. It involves the application of many different bodies of knowledge ranging from abstract concepts to concrete technical know-how. It would be very difficult for the designer to try to resolve all aspects of a design at the same time at the onset of a design problem. When developing a design scheme, designers would start with the basic parameters and layer on additional aspects for consideration as the scheme becomes thought out in more and more detail. When considering each additional aspect, the designer is actually evaluating what has been worked out thus far against a new set of criteria, e.g. structure, statutory requirements, morphological aspects, etc.

The sketch in Figure 4.7 followed the one in Figure 4.6 and was done when the structural implications of the scheme was taken into consideration. After evaluating the structure system for the scheme, the first major change was moving the egg-shaped hall in the harbour back onto dry land and changed its shape to a simpler rectangular shape to avoid complicated construction. The form and composition of the rooms also began to take shape when a structural grid was overlaid onto the scheme. In this sketch all floors of the art school are drawn directly on top of each other to study the relationship between spaces on the different levels.

What is happening here at this stage is that the designer is implicitly evaluating the design scheme with respect to certain aspects that are not previously taken into consideration when adjustments are made in accordance with these newly considered factors. Each new sketch or diagram drawn represents another round of implicit evaluation, i.e. another layer of information/factors is being added onto the design scheme. It is this series of successive implicit evaluations that drives the design development from its rudimentary stage to a detailed and more refined solution.

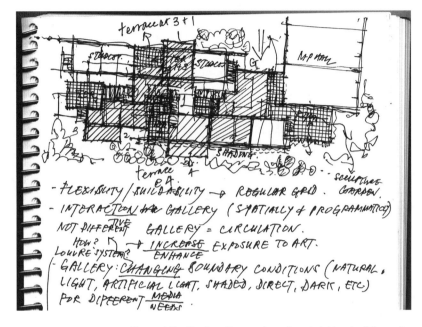

Figure 4.7   Design diagram from the sketchbook of the writer

E. Communication

Any great idea is only as good as how it is conveyed to others; that is, how you communicate your design. There are many forms of communication in architectural design. Drawing sketches is a way of communication with yourself on the development of concepts and ideas. In general, most effort is spent on two types of communication material: presentation material and production material.

Production drawings are used to document the chosen proposal in detail, including all possible physical descriptions and instructions on how it is put together. This document, or in most cases a set or even sets of documents, will be forwarded to the production team, which follows the document step by step to create the design product. These include drawings for general description purposes, such as floor plans (Fig 4.8), elevations and sections, and to provide more detailed information such as detailed drawings. The plan below was developed from the previous sketches shown in this chapter and the similarities between this plan and the sketch in Figure 4.7 are clearly discernible.

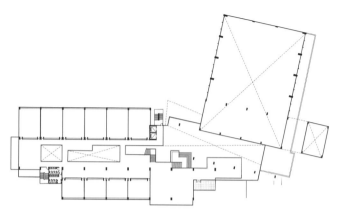

Figure 4.8   First floor plan of the art school proposal

Figure 4.9 is an example of a computer rendering which helps to depict the outlook of the proposed design. In many cases in Hong Kong and China, this is the presentation drawing that the clients are most interested in and often jobs are won or lost based on the quality of this rendering. Other examples of presentation materials are diagrams, study models, computer animation, etc. These mainly serve two purposes: representation of the finished outlook or explanation of the design scheme.

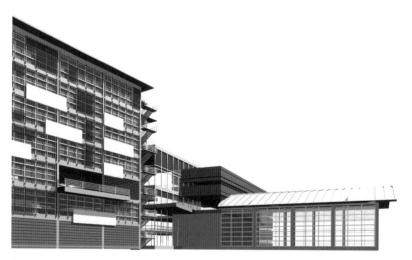

Figure 4.9   Computer rendering of the art school design

4   The Design Loop

In a simple model, the design process is a linear progression of the above five design stages from the inception (formation of problem) to the conclusion (communication of solution) of design. The designer operates in an orderly manner from problem to solution. The designer begins by understanding the problem and proceeds to collect and analyze the various kinds of information available. Once the designer has formed the problem and analyzed the information, he is ready to generate solutions for evaluation. The solution that passes the evaluation is then documented and sent to the production team.

Although most designer would go through all five stages of the design process, most of them concentrated mainly on two activities: forming the problem and generating solutions. They would start by trying to understand the problem, but before adequate information is collected and analyzed, they would attempt to generate possible solutions directly to satisfy some of the more noticeable constraints. But as they proceed and discover that their initial solutions might not satisfy other criteria, they would immediately go back to refining their understanding of the problem and generate new alternatives or modify the previous solution(s).

Designers would jump between problem and solution constantly before stopping at the "final solution" when they ran out of time. It seems that if time is not a factor, they would just keep going on and on between problem and solution to refine their solutions. Through each cycle, the designer would consider new questions — both questions they failed to consider before and questions arising from the interim solutions. And each additional question would change the designer's understanding of the problem and the formation of the problem becomes more and more complex as increasing levels of details were considered. A loop is thus formed between the stages of evaluation and formation where interim solutions are continuously generated (Fig 4.10):

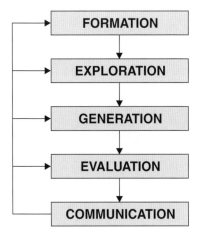

Figure 4.10   The "Design Loop" and the five stages of design

5 Conclusion

The form of the building is not "found" at the end of the above-mentioned process like some treasure being dug up after a long search. Finding form in architecture does not translate quite so literally into this kind of serendipitous discovery. An architect finds form by getting clues and hints from various sources throughout (and back through) the different stages of design that contribute towards the evolution of the design form.

The design form grows from abstract to concrete in accordance with each successive layer of investigation and evaluation that the architect makes in the course of design. This is by no means a hundred-metre dash to the finish line where every single step taken is towards a pre-determined goal. Instead, designing is more like finding one's way out of a labyrinth: not every step may lead you closer to the exit but every step you take, every wrong turn you make, helps you gain a clearer picture of the surrounding and eventually you will find the way out.

References

1. Alexander, C. 1964. *Notes on the Synthesis of Form*, Cambridge: Harvard University Press.

2. Hillier, B. 1996. *Space is the Machine*, Cambridge: Cambridge University Press.

2. Jones, J. C. 1970. *Design Methods*, New York: Van Nostrand Reinhold.

4. Rowe, P. 1987. *Design Thinking*, Cambridge: The MIT Press.

5. Steadman, J. P. 1983. *Architectural Morphology*, London: Pion Limited.

Chapter 5

Hong Kong Architecture: Identities and Prospects — A Discourse on Tradition and Creation

Charlie Q. L. XUE

A recognizable identity is important to any place because of social needs and spatial behaviours. This chapter examines the physical characteristics of Hong Kong architecture, where buildings are characterized by their height, high density and the hilly terrain on which many are erected. As a focal point of East Asia, Hong Kong cannot escape from the international discourse on the dichotomies between government and commercialism; East and West; and globalization and indigenous development. The author discusses how Hong Kong's land constraints, financial situation, building regulations, and design professionals will play an increasingly important role in shaping the city's future architectural identity.

5

Hong Kong Architecture: Identities and Prospects — A Discourse on Tradition and Creation

1 Introduction

It is widely accepted that a recognizable identity for any place is imperative for social needs and for cognitive, spatial and territorial behaviours (Downes and Stea 1977, Lynch 1961). Identity here refers mainly to physical identity, which embodies the societal collective consciousness, technical conditions and other factors. Architectural identities are formed gradually under the social, economic, geographical and climatic influences unique to a place.

Hong Kong is located on the South China coast and nurtured by the humid, warm, subtropical climate. The unique history of the territory, which was ruled by the British government for 155 years and is now under Chinese sovereignty, has given rise to an identity which is quite different from that of its vast hinterland in mainland China.

This chapter will attempt to reveal and analyze the identity of Hong Kong architecture and offer suggestions for maintaining the existing identity and creating new identities in the 21st century. In a wish to provoke debate, the future of Hong Kong architecture is also addressed.

2 Population Pressure and Land Shortage

With a total area of 1,080 sq. km., Hong Kong is only one-fifth the size of greater Shanghai. In 2000, Hong Kong's total population numbered about 7 million, almost twice that of New Zealand. In addition to natural growth,

150 immigrants from mainland China plus others from elsewhere arrive everyday. Continuing immigration has complicated the problem of an increasing population (Ta Kung Pao, 8 Feb 1999; Ming Pao Daily News, 10 Feb 1999).[1] It is estimated that the Hong Kong Special Administra- tive Region (HKSAR) population will reach 8.6 million by 2006, and 10 million by 2010 (HKSAR Planning Department, 2001).

Of Hong Kong's total land area, 40% comprises country parks, islets and landscaped areas. Built-up areas account for about one-third of the territory. Settlement density in such a small city is something of a world record. In areas like Mong Kok and Sham Shui Po, there are on average 10,000 people per hectare. Most days some 700,000 people pass through a small plaza in front of the MTR "Bank Centre" entrance there, which covers no more than a couple of hundred square metres. Public housing estates contain about 2,500 people per hectare, which is double the density of the densest residential areas in Beijing or Shanghai (Figure 5.1).

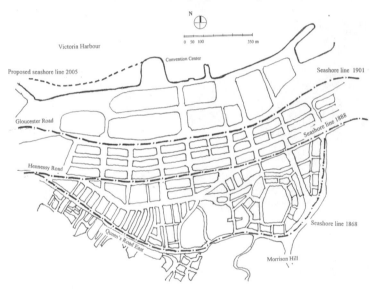

Figure 5.1   The evolution of Wan Chai, Hong Kong Island, through the years

Building regulations play a pivotal role in such a high-rise and high-density environment. Under Hong Kong's building regulations, a plot ratio of 15 and a site coverage of 100% are permitted. These two regulations would be unthinkable in mainland China where the central business areas of Shanghai and Beijing typically allow a plot ratio of 7–9 and never more than 11, with usually 40% greenery coverage. Developers in Hong Kong, on the other hand, tend to fully utilize the development potential suggested in the building regulations, outline zoning plans and lease conditions.

In Hong Kong, land prices sometimes account for 70–80% of the total development cost. Only a high-rise and high-density solution can absorb the astronomical cost of land in the city. The combination of pressure from population growth and a shortage of buildable land means high-rise, high-density development is inevitable. Both public and private housing blocks built in the late 1990s, like those in Ma On Shan and Junk Bay, commonly reach 40 and even 50 storeys. To create open or recreational space, the buildings are usually arranged shoulder to shoulder, with very few voids in between. Wherever you are in an urban area or new town in Hong Kong, you can feel the impact of such a concrete jungle soaring into the sky (Figure 5.2).

Figure 5.2   High density is a problem Hong Kong people have to face

Buildings have to accommodate a variety of functions and good coordination is necessary in the planning, for example, of the separation of pedestrian and vehicular flow. At the same time, building services, retail outlets, eateries, commercial offices, building management offices, flats and parking lots have to be fitted into a small footprint.

Convenience and proximity are two benefits of the highly efficient use of space. For example, Taikoo Place in Quarry Bay is a complex of interconnected office blocks for more than 30,000 workers. The office city has a floor area of more than 456,000 sq m and is linked to the nearby Mass Transit Railway (MTR) station, bus stations and roads. Taikoo Place represents a typical, modern high-rise, high-density solution in Hong Kong. The complex projects an air of sophistication for tenants and passers-by without glaring aesthetics. Its effective, intelligent building services and high-quality volumes are complemented by generous public spaces that are interspersed with artworks and delicate interior settings.

The design of Taikoo Place has also improved the pedestrian areas in its neighbourhood by providing a buffer zone during rush hours. Taikoo Place is in a part of Quarry Bay that used to be industrial in nature, with warehouses, godowns and newspaper plants. The redevelopment has given the area a new image, and in some ways is acting as a catalyst for other redevelopments and the gentrification of Quarry Bay (Figure 5.3).

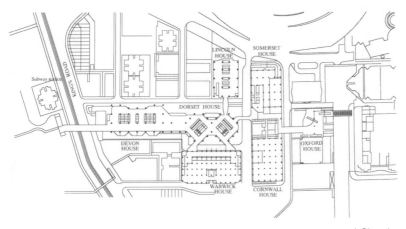

a) Site plan

b) Concourse — typical scene in Hong Kong's office buildings

Figure 5.3   Taikoo Place

Hong Kong architects are used to exploring aesthetics derived from the constraints and pressures of high-rise and high-density living and working. The award-winning Tung Chung Crescent project is one example.

The masterplan of Tung Chung, a new town a mile away from the new Chek Lap Kok international airport, was prepared as part of the airport development. The site of the new town measures about 22.5 hectares. The initial development programme designates 750,000 sq m of space for residential use, 56,000 sq m for retail outlets, 22,000 sq m for hotels and 15,000 sq m for offices. Tung Chung Station Package One is a comprehensive development comprising a total of 13 high-rise residential towers (30–50 storeys each) as well as retail, office and hotel facilities.

Tung Chung Crescent was designed to create a strong, memorable urban form that would give a unique identity to the new town. Residential blocks arranged in a crescent alignment is defined by a generous use of landscaping in front, which becomes the focal point for activities. The distance between the towers is maximized, enabling good views, natural lighting, cross-ventilation and openness for the housing units. The crescent arrangement of the towers also minimizes the number of flats that overlook adjacent developments while maximizing the number with unobstructed sea views. The cascading roofline is a response to the project's site and obtains more view from the favourable direction (Figure 5.4).

Figure 5.4   Tung Chung Crescent

Figure 5.5   Methodist Church,
Wanchai, 1998

In Wan Chai, the old Methodist Church, which was built in the 1930s, was replaced by a new church in 1998 on the same lot. The new church has an office block of more than 20 storeys above it. The entrances to the various parts of the building are well arranged on the ground floor. The podium was topped by a cupola in a symbolic nod to the history of the old church. The delicate and skilful design was in part derived from the constraints of the small site (Figure 5.5).

Most people dream of living in their own house near a beach or in the countryside, but this will only remain a dream for most Hong Kong citizens. Although praised by many scholars, high-rise and high-density housing cannot compare with low-density living in terms of quality. However, for the majority of the 7 million residents in Hong Kong, the latter is an unrealistic alternative.

Many in the West may condemn such high density as inhumane and unacceptable. However, crowded, unauthorized building structures like illegal balconies and rooftop sheds, oversized advertisement boards and night hawkers are often highlighted and praised by others as identifiable characteristics of Hong Kong and treated as major themes for graduation theses by the city's architectural students. Unfortunately, the pitiful life of Hong Kong's working class is often analyzed with an internal logic without reference to standards in the rest of the world and, more importantly, used as an easy formula for developing indeed winning entries for international design competitions.[2] Instead of this romantic view, people should ask: what are the life experiences of ordinary residents? How do they exist on a day-to-day basis in such dilapidated accommodation? Could government and architects do more to improve their living conditions (Figure 5.6)?

a) Student's work, Year 5, Chinese   b) Raymond Chan, Year 3, City University
*   University of Hong Kong, 2001   of Hong Kong, 2001*

Figure 5.6   Students' works take high-density as design topic

Density itself is a physical condition. Through the hands of builders in the past century, it has become part of our tradition. The interpretation of architectural students and professionals represents an effort to recreate this tradition.

3   Southern Chinese Origins

Before and during British rule, the building type adopted for housing most local Chinese was derived from southern China. Examples include the walled villages in the New Territories and buildings with colonnades in Sheung Wan, Hong Kong Island, and Yau Ma Tei, Kowloon. The colonialists also brought their own architectural styles, which were then adapted for the humid climate, local materials and labour skills. Typical colonial buildings in Hong Kong include the main hall of the University of Hong Kong (1907), the former Flagstaff House (now the Tea Museum), the former Kowloon train station in Tsim Sha Tsui, and the former Supreme Court (now home of the Legislative Council) in Central.

From the turn of the 20th century to 1950, the Chinese merchants and developers started to build villas and tenement houses (*tang lou*) in urban Hong Kong. Tenement houses, as defined in the building regulations, are buildings in which all or most flats are occupied by more than one household. The layout of a typical *tang lou* is influenced by the vernacular buildings of southern China (Cheung 2000). Buildings of the 1930s and 1940s that show an obvious Chinese influence are loosely called "*tang lou*" too. The unique layout, decoration and materials of tang lou could be compared to Shanghai's *lilong* (terraced-lane houses) and Beijing's courtyard houses (Figure 5.7).

Figure 5.7   Old golry becomes modern burden (Lui Seng Chun)

Tang lou or tenement houses often fell into a dilapidated state after half a century of use. They are high in density and rundown, and are usually found in Hong Kong's urban areas. These houses are generally overcrowded and feature complicated owner-tenant arrangements, cluttered illegal add-ons such as rooftop sheds, advertisement boards and other temporary structures, making them an urban renewal headache. Kowloon Walled City was romanticised by cameras and its demolition was lamented by people with a misled appreciation of makeshift architecture in developing economies. Although it was demolished, old urban areas in Wan Chai, Yau Ma Tei, Mong Kok, Sham Shui Po and others are still steadily decaying — Kowloon Walled City is actually everywhere (Figure 5.8)!

As *tang lou* and other building forms associated with the mainland or southern China are usually old structures that were perceived as backward, building forms originating from southern China almost disappeared from new construction. A few buildings tried to embody the style of orthodox "Chinese" architecture, even though they do not form the mainstream in Hong Kong architecture and do not even represent a strong trend. Chi Lin Nunnery (1998), the Hong Kong Museum of

Figure 5.8   Kowloon Walled City is everywhere

a) Mong Kok

b) Tai Kok Tsui   *c) Sham Shui Po*

Figure 5.9   Tendency to "speak Chinese"

a) Chi Lin Nunnery

b) Hong Kong Heritage Museum

c) Lingnan University

Heritage (2000), Lingnan University Campus (1996) and some other new buildings show a deliberate attempt to imitate the language and forms of ancient Chinese architecture. However, the quasi-Chinese style of the Heritage Museum is quite unsatisfactory (Figure 5.9).

Except for these very special building types, there seems no justification for most of the buildings to adopt the "pure" Chinese vocabulary. Actually, there are quite a number of advantages to traditional southern Chinese architecture, in terms of, for example, ventilation; the use of local, inexpensive materials; intimacy etc. How they can be applied to high-rise, high density buildings is the question for Hong Kong architects.

4 Politics Giving Way to Commercialism

Hong Kong is not a major centre of political activity like London or Beijing. It has no grand buildings for the sole purpose of parliamentary debate or the workings of government, such as London's Houses of Parliament (1865) and New Delhi's Governor's House (1912–1930). In Hong Kong, the political realm is represented by the Legislative Council Building (formerly the Supreme Court, built in 1912) and the former Governor's House on Upper Albert Road. These two buildings are dwarfed by the surrounding commercial buildings. Government buildings which were built later all look modest, both in terms of shape and location, and these buildings reflect the small role of the government as opposed to market forces in Hong Kong. Banking and financial services are more representative of contemporary Hong Kong and their iconic buildings likewise tend to dominate the cityscape.

There are usually a few glamorous' districts in every city that add a little colour and pace to the local scene. Hongkong Land, an old English property development company, built a series of magnificent edifices on the Central waterfront that were mostly designed by the similarly long-established practice of Palmer & Turner (now P& T Group). The buildings were conceived according to their function, current technology and modern aesthetics, completely free of any obligation to look to traditional Chinese architecture for inspiration (P & T Group, 1998). Hong Kong Bank (1935, 1985), Bank of China (1948, 1989), Prince's Building, Hong Kong Club, Mandarin Hotel, Jardine House, Exchange Square, the Landmark and Entertainment Building could almost be arranged into a timeline tracing Hong Kong's evolution from a small, dirty harbour in the Orient into a major financial metropolis in Asia and indeed the world (Figure 5.10).

a) Elevation

b) Lobby

Figure 5.10   Entertainment building, Central

Figure 5.11   Airport in construction, 1996

The construction of the new Hong Kong Bank building (1985), the Bank of China (1989), Chek Lap Kok International Airport (1998), Airport Express Kowloon Station (1998) and KCRC Hung Hom Station (1998) gave the world — or at least the global community of architects — a taste of the city's sensitive response to available industrial resources, sophistication in the use of space, volume and technology, and aesthetic freshness (Figure 5.11).

These buildings embody, to a certain extent, the identities of modern Hong Kong: an active and sophisticated commercial and financial centre not only of Asia but of the world. These prominent buildings can be framed more in the discourse of globalization than of regionalism as they were developed by giant corporations driven by competition and pride. The huge profits that could be derived from real estate development was also an incentive for many.

The glamorous commercial buildings were mostly designed by overseas architects. For example, the Hong Kong Convention & Exhibition Centre Extension, the venue of the 1997 handover ceremony, was conceived by an American. With a strong colonial past, the city still considers Western entities to be superior, and Hong Kong architects, arguably as talented as their foreign counterparts, have not really been put in the spotlight (Figure 5.12).

Figure 5.12   Convention Centre in Wanchai

Hong Kong has inherited some British traditions in building procurement, system of management, procedures, design approaches, building technologies and materials. Hong Kong architects however do not share the same status as their Western counterparts. In the city, architects are not seen as noble, artistic creators but as cogs in the wheel of property economics — in particular, the speculative activity of property developers and investors driven by greed. The colonial administration only envisioned the territory as a trading outpost in the Far East, and hence never thoroughly planned for its long-term future till the 1970s (Xue, C. and Manuel, K. 2001). The laissez faire policy was successful in some way but it was adopted at the expense of long-term central planning.

The situation improved slightly after the 1980s, but the issue was never completely addressed. The post-handover SAR government is still in a transitional stage, and too many urgent problems associated with housing the homeless and the elderly and supporting low-income families are deemed more important than cultural and architectural identities. While this may be true, physical improvement and identity may not necessarily conflict with the need to address these problems.

5 Globalization versus Indigenous Development

From 1842 until the handover, Hong Kong was a British colony and Britain's outpost in the Far East. Hong Kong has remained an international financial and economic centre after China regained the territory in July 1997. Hong Kong is an example of an Eastern city that has successfully absorbed the advanced technical and managerial techniques of Western countries.

In the last 20 years of the 20th century, there was a heated debate in mainland China concerning "national architecture". Hong Kong was more or less left out of the discussion, and most professionals in Hong Kong never even got wind of the debate engrossing their mainland counterparts.

With the gradual rise of financial and economic power in Asia, Hong Kong began to export its culture as well as services to its vast hinterland on the mainland. In the early 1980s, Hong Kong architects began to design luxury hotels, sumptuous houses and modern office towers in major Chinese cities like Beijing, Shanghai, Chongqing and especially Guangzhou and Shenzhen, further imposing its influence on building names, management styles, land leasing methods and construction systems. A "Hong Kong tide" almost overwhelmed the local culture, so much so that the practice of "learning from the West" on the mainland became "learning from Hong Kong" (Figure 5.13).

During its development over the past 160 years, Hong Kong architecture has evolved an identity which is neither exclusively Chinese nor exclusively British, but something in between. English, as the colony's official language, has evolved into the "Honglish" or "Cantoglish" heard on Hong Kong streets today. A similar analogy can be applied to its architecture.

*Figure 5.13   Hong Kong will
continue to play a part
in "colonizing China"*

Figure 5.14   A generic city?

*Figure 5.15   Rocco Yim's design,
Bamboo Pavilion, 2000*

In common with many other major cities around the world, there is a notable influx of an "international" style of architecture that is supranational. Due to the spread of globalization, Hong Kong has become more and more like a member of the global village. The past 30 years have witnessed an exponential growth in the number of overseas investment and trading companies in Hong Kong, which has been, intentionally or otherwise, pushed into the ranks of "generic cities" — a term coined by pioneering architect and theorist Rem Koolhaas — together with cities like Singapore, Tokyo and Bangkok. In such "generic" Asian cities, airports, highways, and public spaces are characterised by forms which are simmered down and superimposed on a locale without consideration for distinctive identities, genius loci, the human dimension or historical continuity (Koolhass and Mau 1995) (Figure 5.14).

Another trend has emerged as globalization gathers pace. Although there may be some mainland Chinese or foreign influence, Hong Kong's culture has largely gone its own way. People in the middle class, the working class, minorities and mainland immigrants have their own values and way of life. Those born in Hong Kong have driven much of the evolution of the city's characteristics while the economic miracle of the past 30 years provide fertile ground for a sense of belonging (Ng 1998). Residents share a common culture and sense of achievement and often distinguish between ideas developed domestically and overseas.

Architecture in the city is slowly responding to the rise of a sense of belonging even though the majority of major infrastructure projects are still driven by the commercial interests mentioned above. The Central Library in Causeway Bay shows a strong intention to mix West and East in its collage of different vocabularies. Perhaps burdened by heavy expectation of its symbolism, the building suffers from a chaotic elevation design and attracts intensive criticism in the city.

Some architects are exploring an architecture which is more adapted to local conditions. For example, Anthony Ng's designs for Verbena Heights and Tung Chung Crescent respond to the hot and humid climate. Rocco Yim's design for the Kowloon Park entrance (1986) defines and punctuates the fabulous Tsim Sha Tsui part of Nathan Road. His design for the Peninsula Hotel extension (1994) fully respects the 1930s British style of the existing building while creating a grand new building overlooking the harbour. Yim's design of a bamboo pavilion for the "Hong Kong — Berlin Cultural Festival" in 2000 evoked a simple yet dynamic spirit that belongs to both Hong Kong and China (Figure 5.15).

Hilly terrain and uneven ground are special topographical features of Hong Kong. They put a constraint on construction but can also inspire unique designs. They present a challenge as well as an opportunity that Hong Kong architects have learnt to grasp. Examples include the Hong

Figure 5.16   Hilly buildings in Hong Kong: a) Campus of Hong Kong University of Science and Technology

b) Graduate Students' Hall, University of Hong Kong, 1999

Kong University of Science and Technology campus; Phase IV of Graduate House, University of Hong Kong; and Jockey Club Environmental Building in Kowloon Tong, to name but a few. These buildings were skilfully designed to suit the level differences (Figure 5.16).

Globalization speeds up communication across the world and, in doing so, can bring about good and bad effects. The familiarity and convenience gained are offset by the elimination of differences, resulting in the emergence of global homogeneity that tends to threaten the existence of local forms. However, local material conditions, topography, climate and, sometimes, people's habits will remain relatively unchanged. Perhaps more time is needed before the whole of Hong Kong's society embraces a common identity and sense of belonging.

6   Conclusion — Prospects for Hong Kong Architecture

Some of the suggestions mentioned in the above description will not be repeated in this section even though, for the sake of coherence, they should be included in the conclusion. Based on the above discussion and with a wish for a better tomorrow, it is predicted that the following issues will influence Hong Kong's architectural identity in the future:

6.1   Increase in land constraints

If the current rate of immigration growth continues, Hong Kong's population will total 8.6 million by 2006. That means housing must be provided for an extra 1.6 million people (about 53,000 families). Based on the current density in public housing (2,500 persons/hectare), about 600 hectares of new land will be needed. To enhance living quality and lower the density, more land will be needed. Reclamation and urban renewal plans currently approved by the Government cannot meet this demand, however, due to ongoing disputes over harbour reclamation and encroachment on suburban parks. From a layman's point of view, the islets seem to provide some development potential, but developers may be intimidated by the high infrastructure cost. A 300-storey, "bionic" tower was recently proposed for Shanghai (*Ming Pao*, 26 Feb 2001), and it is inevitable that Hong Kong will also have to consider the feasibility of "hyperdensity" in the future.

6.2   Need for revision of outdated building regulations

There is no doubt that building regulations help to mould the modern architectural identity of Hong Kong, but the regulations also impede healthy development to a certain degree. After protracted appeals and severe criticisms, the government eventually exempted balconies from the calculation of gross floor areas (but the government asked for premiums as compensation) in 2001, to encourage the installation of balconies and other more environmentally-friendly features. Building regulations have a statutory role to play, but they impose limits that confine buildings to a particular place and a particular time. In the past 50 years, Hong Kong's building regulations have always lagged behind societal demands and changes. Even today, there are still many unreasonable regulations that conflict with users' behaviour, normal building functions and common sense, as indicated by several research studies (Hui and Chan 2000, W. S. Wong 2000). A thorough overhaul of building regulations is urgently needed.

6.3   Financial condition may worsen

After a 30-year boom, Hong Kong was severely hit by the 1998 economic downturn in Asia. The city's building industry was quiet during 1998 and 1999. Although signs of a recovery emerged in 2000, the economy plummeted again severely in the global recession of 2001. As mentioned previously, Hong Kong's existing architectural landmarks were produced mainly by strong financial players. Excellent designs can be achieved despite the constraint of cost, but generally speaking, deep pockets are indispensable to creations of a high standard. Less money means less incentive to construct new buildings. With China's entry into the World Trade Organization and the resulting opening up of access to its ports and currency, Hong Kong will gradually lose its influence and advantage as the gateway to China. In light of these factors, Hong Kong architects should embrace the wider markets present in the Pearl River Delta, southern China, and indeed the whole of the mainland and Southeast Asia.

6.4   Accelerate China's "colonization"

Before the 1980s, Beijing, Shanghai and Guangzhou each displayed distinctive architectural styles. Guangzhou, for example, developed its "Lingnan School", which attracted the attention of China's architectural circles.[3] As previously mentioned, Hong Kong has made its presence felt through its architects' work in various parts of the mainland, and they should endeavour to continue doing so. Although Shanghai, with its strong culture and history, is undergoing a renaissance in economic and social importance, Hong Kong should at least try to keep a leading role in southern China.

6.5   Mindsets need to change

The term "creation" is often used instead of "design" to describe many buildings in mainland China. In contrast, Hong Kong architects do not normally think of their buildings as artistic creations; they put greater emphasis on buildings "designed" carefully and professionally, in strict accordance with proper procurement procedures.

Buildings in Hong Kong are also regarded as a commodity subject to speculation, hence contemplation of them as cultural creations or structure with unique identities is often rare or weak. Hong Kong does not have a culture that encourages the creation of identities in architectural design, which is evident in the lack of architectural competitions, criticism, awards and social awareness.

The HKSAR's Chief Executive Tung Chee Hwa pledged to turn Hong Kong into an international metropolis that can be compared to London and New York *(Policy Address* 1999). But does Hong Kong have any professional architectural magazine that can rival in quality the U.K.'s *Architectural Review* or the U.S.'s *Architectural Record*, or those countries' academic journals (e.g., *Journal of Architectural and Planning Research*; *Urban Design International*), or even Beijing's *Architectural Journal and Architect* magazine? As the territory's professional body for architects, the Hong Kong Institute of Architects has a social obligation to foster the development of an environment conducive to great architectural achievements.

The majority of Hong Kong's architectural design professionals received their education in the territory or overseas, especially the U.K.; elsewhere in Europe; Australia, Canada and the U.S. At the same time, many techniques were brought to Hong Kong by foreign consultancies and localised over the years. Graduates and professionals from mainland China have contributed and continue to actively contribute to Hong Kong's building industry too. Thus, the contribution made to Hong Kong architecture by people from overseas should not be underestimated.

When more and more Hong Kong architects and building professionals work in China, a reverse flow could occur whereby more and more mainland professionals will come to Hong Kong. When investments start flowing into Hong Kong from the mainland, the city's investors will begin to consider engaging (their) design partners in Beijing, Shanghai or elsewhere in mainland China for projects in Hong

Kong. Mainland building contractor China Construction entered the Hong Kong market some 15 years ago and established China Overseas Construction in the city. A similar development may happen in the realm of architecture in the near future. The emergence of mainland designers in the territory will also bring new ideas and diversity to Hong Kong's current architectural thinking.

The above discussion concerns the evolution of an identifiable Hong Kong architecture. Improvements in procedures and systems may help foster the emergence of strong architectural identities which may one day mark Hong Kong out from the crowd in Asia and the world.

The author thanks the constructive advice from Professor Bill Lim. Gratitude is also due to Chen Xi, whose photographs are used for the cover of this book as well as the title page of this chapter. This chapter is part of a study supported by City University of Hong Kong, Project No. 9030794, and Research Grant Council, Hong Kong Special Administrative Region Government, Project No. CityU 1053/00H.

Notes

1. The controversy over the right of abode of children of married or unmarried Hong Kong citizens was triggered by the Hong Kong Supreme Court's decision in one case in January 1999. If the children of unmarried Hong Kong citizens are to be granted right of abode, there would be a population increase of 1.6 million over four years. The particular case was sent to the Court of Final Appeal and was interpreted by the People's Congress of Central Government in June 1999. See *South China Morning Post*, 20 Jan–10 Feb, and May–June, 1999.

2. Students from the University of Hong Kong have won numerous international students' competitions, in particular those organized by the American Collegiate Society of Architecture, from 1990 onwards. Most winning schemes involved conversions or reuse of old and dilapidated sites, for example, Tai O fishing village, Sheung Wan, etc.

3. Lingnan loosely refers to the provinces of Guangdong, Guangxi and perhaps part of Fujian. Because of its distance from the political centre, the natural, non-orthodox and non-official style easily found root in this area. In the 20th century, Lingnan nurtured its unique schools of painting, literature and architecture.

References

1. Chan, Man Hung. March 2001. "Ru shi hou xiang gang jiang bin jinji zhuangxin wei ji." (Hong Kong facing crisis of transformation after China enters WTO). Hong Kong: *Ming Pao Monthly*, No. 3.

2. Cheung, Ferdinand K. H. 2000. "Tenement buildings: in light of their origin", *Hong Kong Institute of Architects Journal* 24 (2nd Quarter): 78–87.

3. Department of Architecture, Hong Kong University. 1999. *Measured Drawings*, Vol. I and II., Hong Kong: Pace Publishing Ltd.

4. Downes, R. M. and Stea, D. 1977. *Maps in Minds: Reflections on Cognitive Mapping*. New York: Harper and Rowe.

5. Eu, Geoffrey. 1995. *Insight Guides: Hong Kong*. Singapore: APA Publications.

6. Ganesan, S. 2000. *Employment, technology and construction development: with case studies in Asia and China*, Aldershot, U.K.: Ashgate.

7. Ganesan, S. and Lau, Stephen. 2000. "Urban challenges in Hong Kong: future directions for design." *Proceedings of International Conference, Megacities 2000*. Vol. 1. University of Hong Kong.

8. Hong Kong Institute of Architects. 1999–2000. *Annual Exhibition Publication*.

9. Hong Kong Institute of Architects. 2000. "Taikoo Place, Project news." *Hong Kong Institute of Architects Journal*. 25 (3rd Quarter).

10. Hui, Eddie C. M. and Chan, Edwin H. W. 2000. "Impacts of government land leases conditions on private housing market in Hong Kong," *Hong Kong Institute of Architects Journal*. 25 (3rd Quarter).

11. Koolhass, Rem and Mau, Bruce. 1995. *S, M, L, XL*, New York: The Monacelli Press.

12. Lung, P. Y. 1991. *Xianggang gujin jianzhu (The ancient and today's Hong Kong architecture)*. Hong Kong: Joint Publication.

13. Lynch, K. 1961. *The Image of the City*, Cambridge: MIT Press.

14. Manuel, Kevin K. and Xue, Charlie Q. L. 1999. "The civic space of urban Hong Kong after July 1, 1997 — arguments and strategies in public space design", *Academic Treatise of the XXth UIA Congress of World Architects*, Beijing, June 1999.

15. *Mingpao Daily News*, Hong Kong, February 8–20, 1999, February 1–28, 2001.

16. Moore, Steven A. 2001. "Technology, Place, and the Nonmodern Thesis," *Journal of Architectural Education*. Cambridge: MIT Press. 54(3): 130–139.

17. Muramatsu, Shin. 1995. *Chokyu Asia Modern* (Modern Asian Cities). Tokyo: Kaji Ma Institute Publishing Co.

18. Ng, Jun Hung. 1998. "Xunzhao xianggang ben tu yi shi" (Finding the local ideology). *Ming Pao Monthly*. 3 (1988).

19. P & T Group. 1998. *P & T Group*. Hong Kong: Pace Publishing Ltd.

20. Planning Department. 1990. *Hong Kong Planning Standards and Guidelines,* Chapter 4. Hong Kong Special Administrative Region.

21. _____. 2001. *Hong Kong 2030 — Planning vision and strategy.* Hong Kong Special Administrative Region.

22. *Ta Kung Pao.* 8–10 February, 1999.

23. Tung, Chee-hwa. 1999. *Policy Address 1999.* Hong Kong Special Administrative Region.

24. Tsui, Edward. February 2000. "Hyperdensity." *Proceedings of International Conference Megacities 2000.* Vol. 1. University of Hong Kong.

25. Wang, Weijun. 2000. "Writing between the Orientals: Asian city identity and design education", *Conference Proceedings*, Continuity & Innovation — Chinese Conference on Architectural Education, Hong Kong, August 2000.

26. Wong, Wah Sang. 1999. "Legislative control for quality buildings", *Hong Kong Institute of Architects Journal.* 22 (4th Quarter): 20–32.

27. Xue, Charlie Q. L. 1997. "Hong Kong architecture: the hopelessness and hopeness." *The Architect* (Beijing). 76: 97–101.

28. _____. 1998. "Urban architecture of Hong Kong." *The Architect*, Beijing. 80: 39–49.

29. _____. 1999. *Building Practice in China.* Hong Kong: Pace Publishing Ltd.

30. _____. 2000. "Ten celebrities of Hong Kong and Shanghai." *Hong Kong Institute of Architects Journal.* 26 (4th Quarter).

31. _____. 2001. *Contemplation on Architecture.* Hong Kong: Pace Publishing Ltd.

32. _____. 2002. "Organic Renewal of Housing in Old City Area: A Case Study of Hong Kong." *International Journal of Housing Science and Its Application*, 26 (1): 27–28.

33. Xue, Charlie Q. L. and Manuel, Kevin K. 2001. "The quest for better public space, a critical review of urban Hong Kong." In Miao, P. (ed.) *Public Places of Asia Pacific Cities.* The Netherlands: Kluwer Academic Publishers. 171–190.

34. Xue, Charlie Q. L., Manuel, Kevin K. and Chung, Rex H. Y. 2001. "Public space in the derelict city area — a case study of Mong Kok, Hong Kong." *Urban Design International.* 6 (1): 15–31.

35. Xue, Charlie Q. L. and Chung, Rex H. Y. 2001. "Organic development — an alternative approach for urban renewal in Hong Kong", *Hong Kong Institute of Architects Journal.* 28 (2nd Quarter): 54–59, and 29 (3rd Quarter): 52–57.

Chapter 6

Analysis and Design of Structures Using Spreadsheets

Jackson KONG

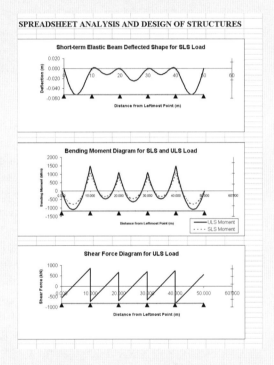

The purpose of this chapter is two-fold, namely, (1) to provide an overview of the relationship between a real structure and its computer model; and (2) to demonstrate the details of computer modelling through an example of continuous concrete beam design using spreadsheets. This example also serves to illustrate that spreadsheets are user-friendly and extremely powerful as a supplementary computer tool for practical structural design. For readers who are interested in learning how to apply spreadsheets to solve specific problems, the sample spreadsheets shown in this chapter and other spreadsheets for various structural design problems are available from the author's homepage (http://personal.cityu.edu.hk/~bsjkong).

6

Analysis and Design of Structures Using Spreadsheets

1 Computer-aided Structural Analysis and Design

Computer-aided analysis and design (CAAD) seems to be the norm in the current practice of structural engineering, particularly in the analysis and design of tall buildings, multi-span bridges and other complicated structures. Commercial software and in-house computer programmes are both commonly used in structural design offices. Most of these sophisticated structural analysis and design software are developed using the finite element method (FEM) (Cheung, Lo and Leung 1996) and are applicable to a wide range of structural problems. In addition to easy-to-use graphical interface, recent developments in these software packages include linkages with other computer tools such as CAD, MS Word, Excel, MathCAD and with various country codes of practice.

Spreadsheets, on the other hand, are well known and well-used for management purposes. They are also commonly used to supplement commercial software in handling input and output data files. However, they are not perhaps being exploited as much as they should be in structural design, where they have tremendous ability to speed up the design process by utilizing a lot of special built-in features currently available. Whether spreadsheets or commercial software are used for a specific structural problem, however, it is important for the designer to understand the general behaviour of the structure to be analyzed, the computer models to be used and the relatership between them. In this respect, it is the purpose of this chapter to provide an overview of the relationship between the behaviour of a real structure and its computer model, and to demonstrate the versatility of applications spreadsheets in structural engineering through an example of continuous concrete beam design.

2   A Real Structure and Its Computer Model

In computer-aided structural analysis and design, a real structure is usually modelled first as an assemblage of *finite elements*, thus creating a mathematical model (i.e. an FEM model) for analysis. Very often, different types of finite elements are used to model different parts of a structure, depending on its structural behaviour and the modelling approach. Prior to the subsequent design of individual structural components (beams, columns, slabs, walls etc.), the computed results are checked against (1) the behaviour of the real structure as a whole and (2) appropriate checking and verification models (Macleod 1988). If the results are reasonable, the design of each individual structural component is then carried out based on the computed forces and moments according to various codes of practice. It is therefore obvious that an understanding of the relationship between the mathematical model (FEM model) and the real behaviour of the structure to be analyzed are crucial in computer-aided structural design (Macleod 1990; Meyer 1987). A summary of this relationship is given below:

2.1   Loading

A real structure is usually subject to various types of loads during its service life, including (1) static load due to the structure's own weight, superimposed load and differential settlement of foundation; (2) dynamic load due to traffic, impact from traffic, wind and earthquake; and (3) time-dependent load due to creep and shrinkage of concrete and temperature variation. Except for the structure's own weight, the loads are complex and cannot be easily and accurately defined. Fortunately, in most situations, the complexity (such as the transient, dynamic and random nature) of these loads can be "simplified" and modelled as "static" loads for analysis and design. Depending on the type of load, the definition of loading may differ from country to country.

Despite the above simplification, the design of a real structure often requires the consideration of a large number of possible loading cases and combinations and the most critical load combination needs to be sorted out for the design of an individual structural component. To minimize the number of load cases and combinations to be considered in the analysis and design, a thorough understanding of the behaviour of the structure is necessary.

2.2   Material behaviour

Although non-linear analysis techniques and software are now available, most FEM models of tall buildings and bridges are still based on the assumptions that the materials are linear elastic and the structural deformation is negligible when compared with the original structural configuration. Such assumptions imply that the force-deformation relationship of the structure is linear and the deformation is reversible upon release of loading. In addition, the force and deformation of the structure under various loading cases can be simply added together to generate a combined "response" of the structure under various load combinations (i.e. the principle of superposition applies). However, it is noteworthy that, despite the analysis of the structure being based on the assumption of linear elasticity, the design of an individual structural element adopts the so-called ultimate limit state theory (Kong and Evans 1987), which is based on the actual realistic behaviour of the materials and its modes of failure.

2.3   Boundary conditions

Boundary conditions (or supporting conditions) of a structure are often difficult to model accurately. Perfectly pinned or fixed conditions at the base of columns or walls, as usually adopted for analysis, are not easy to achieve in practice, on site. Detailing of the connections should therefore be made as closely compatible with the design assumptions as possible, in order not to generate any undesired effects on the structure.

2.4   Equilibrium

For a real structure in equilibrium under the action of a system of forces and boundary conditions, the basic conditions of equilibrium should be satisfied within the domain of the structure. However, as mentioned previously, most of the structural analysis and design software packages are based on the finite element method upon which an FEM model is established. In this model, displacements and stresses within each finite element are only an approximation of the real structure and equilibrium is achieved only in an "integral" sense (Zienkiewicz and Taylor 1989). In other words, equilibrium will not be exactly satisfied at each point within the FEM model unless the approximation matches exactly with the real structural behaviour, which is rarely the case. From a practical point of view, in order to establish confidence in the FEM model, it is therefore essential, upon completion of the analysis, to carry out spot checks to determine that equilibrium is at least approximately satisfied at a few critical locations.

2.5   Compatibility of deformation

Deformation between various connected structural components, such as beams and columns, slabs and walls is usually compatible and continuous in the real structure. However, this compatibility of deformation does not necessarily exist in the FEM model, depending on the types of finite elements used to model the different structural components. Accuracy and convergence of the computed results cannot be guaranteed when different finite elements with incompatible deformation are used to model each of the connected components. An understanding of the behaviour of the structure as a whole (as well as its individual components) and the FEM model is indispensable to proper modelling of structures.

2.6   Construction stages

In addition to the above consideration of the structure during its service life, it is worth noting that, depending on the type of structure, construction methods and the construction sequence, it is sometimes necessary to take into consideration the gradual "evolution" of the structure during construction. A typical example that requires such consideration would be a continuous prestressed concrete bridge.

3   Design of Continuous Reinforced Concrete Beams and One-way Slabs Using Spreadsheets

In this section, the versatility and application of spreadsheets in structure enginnering are demonstrated through a practical design of a continuous concrete beam. The spreadsheet in the following section is designed to help young engineers test their design through simple modeling and to visualize and understand how materials and geometry of a structure change during the process.

3.1   Theory

Prior to the analysis of any structural systems, the sectional properties of structural members are computed. Very often, simple formulae can only be easily found for simple cross sections, while for complicated cross sections, commercial or in-house software packages are used. In the event of the latter, a large number of input data files need to be prepared. However, based on first principles and simple engineering mathematics, sectional properties of different cross-sections can be easily computed using the framework described below.

Sectional properties

By means of Green's Theorem, it can be shown that the following surface integrals of domain R can be reduced to line integrals over its closed boundary:

$$\iint dx.\,dy \quad = \quad (\int x\,dy - \int y\,dx)\,/\,2$$
$$\iint x\,dx.\,dy \quad = \quad (\int x^2\,dy - \int x\,y\,dx\,)\,/\,3$$
$$\iint y\,dx.\,dy \quad = \quad (-\int y^2\,dx + \int x\,y\,dy\,)\,/\,3$$
$$\iint x^2\,dx.\,dy \quad = \quad (\int x^3\,dy - \int x^2\,y\,dx\,)\,/\,4$$
$$\iint y^2\,dx.\,dy \quad = \quad (-\int y^3\,dx + \int x\,y^2\,dy\,)\,/\,4$$

Here all line integrals are carried out in the positive direction of integration around a closed curve.

Assuming that the boundary of the cross-section is divided into n points (in counter-clockwise direction), then the sectional properties can be evaluated using numerical integration, i.e.

Area of cross section $\quad$ A $\;=\; \iint dx.\,dy$
$$= \Sigma\,(x_i\,y_{i+1} - x_{i+1}y_i)$$

Distance of the neutral axis Ax $\;=\; \iint x\,dx.dy$
$$= \Sigma\,(x_i\,y_{i+1} - x_{i+1}y_i)\,(\,x_i + x_{i+1})\,/\,6$$
$$Ay \;=\; \iint y\,dx.\,dy$$
$$= \Sigma\,(x_i\,y_{i+1} - x_{i+1}y_i)\,(\,y_i + y_{i+1})\,/\,6$$

Second moment of inertia $\quad I_x \;=\; \iint y^2\,dx.\,dy$
$$= \Sigma\,(x_i\,y_{i+1} - x_{i+1}y_i)\,(\,y_i^2 + y_i\,y_{i+1} + y_{i+1}^2)\,/\,12$$
$$I_y \;=\; \iint y^2\,dx.\,dy$$
$$= \Sigma\,(x_i\,y_{i+1} - x_{i+1}y_i)\,(\,x_i^2 + x_i\,x_{i+1} + x_{i+1}^2)\,/\,12$$

As a vehicle to demonstrate the application of the aforesaid numerical approach, the sectional properties of an I-section are computed using the spreadsheet as shown in Figure 6.1. Using exactly the same framework, the sectional properties of a box-section can also be easily determined by changing only the coordinates of the points around the boundary.

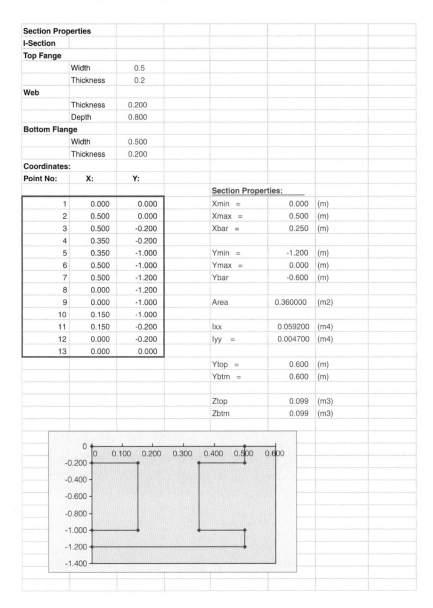

Section Properties		
I-Section		
Top Fange		
	Width	0.5
	Thickness	0.2
Web		
	Thickness	0.200
	Depth	0.800
Bottom Flange		
	Width	0.500
	Thickness	0.200
Coordinates:		
Point No:	**X:**	**Y:**

Section Properties:

Xmin =	0.000	(m)
Xmax =	0.500	(m)
Xbar =	0.250	(m)
Ymin =	-1.200	(m)
Ymax =	0.000	(m)
Ybar	-0.600	(m)
Area	0.360000	(m2)
Ixx	0.059200	(m4)
Iyy =	0.004700	(m4)
Ytop =	0.600	(m)
Ybtm =	0.600	(m)
Ztop	0.099	(m3)
Zbtm	0.099	(m3)

Point No	X:	Y:
1	0.000	0.000
2	0.500	0.000
3	0.500	-0.200
4	0.350	-0.200
5	0.350	-1.000
6	0.500	-1.000
7	0.500	-1.200
8	0.000	-1.200
9	0.000	-1.000
10	0.150	-1.000
11	0.150	-0.200
12	0.000	-0.200
13	0.000	0.000

Figure 6.1   A screen dump from a spreadsheet for computing sectional properties of various cross sections.

Analysis of continuous beams

A continuous concrete beam is analyzed using the finite element method. Each span of the continuous beam is modelled by a so-called beam-type finite element, and the force-deformation relationship is represented by the following 4×4 finite element stiffness matrix:

$$\begin{bmatrix} 12EI/L^3 & 6EI/L^2 & -12EI/L^3 & 6EI/L^2 \\ & 4EI/L & -6EI/L^2 & 2EI/L \\ Symmetric & & 12EI/L^3 & -6EI/L^2 \\ & & & 4EI/L \end{bmatrix}$$

By enforcing compatibility of deformation between adjacent beam elements, the stiffness matrices of all beam elements are assembled into the final global stiffness matrix [K]. Applied loads due to the beam's own weight and superimposed loads are modelled as uniformly distributed load or point loads. Load vectors are then generated and assembled to form the global load vector {F}. Equilibrium of the structure is then represented by the matrix equation:

$$[K]\{d\} = \{F\}$$

By imposing the boundary (support) conditions, the above equation can be easily solved:

$$\Rightarrow \{d\} = [K]^{-1}\{F\}$$

Here {d} is the displacement vector computed by using the built-in function of matrix inversion of the spreadsheet. Deflection, bending moments and shear forces can be subsequently calculated. Reinforcements are then designed in accordance with the relevant codes of practice.

3.2   Implementation

Figure 6.2 shows a screen dump from the spreadsheet for the basic data input of the continuous concrete beam. Geometric parameters including the number of spans, span lengths, support conditions and material properties are entered therein. The "CONFIRM INPUT" button illustrated has a macro (a specially written subroutine) assigned to it which automatically updates all relevant data before analysis. Loading and beam sizes are then entered into another spreadsheet, as shown in Figure 6.3. Based on the procedures as outlined in the previous section, the response of the structure, in terms of deflection, bending moment and shear force distribution, is calculated and instantly plotted, thus facilitating the visualization and understanding of its behaviour. (A screen dump from the spreadsheet for calculating the inverse of the global stiffness matrix is also

Beam Mark	MB1		
Span Basic Info			
Total no. of Spans (Including Cantilever, max 5)	5		
First Span is Cantilever (Y/N)	N		
Last Span is Cantilever (Y/N)	N		
Span Length			
Span Length #1	10.000	m	
Span Length #2	10.000	m	
Span Length #3	10.000	m	
Span Length #4	10.000	m	
Span Length #5	10.000	m	
Material Parameters			
fy (N/mm^2)	460		
fcu (N/mm^2)	40		
Eshort (kN/m^2)	2.40E+07		
Elong (kN/m^2)	1.20E+07		CONFIRM INPUT
Esteel (kN/m^2)	2.05E+08		
creep factor	1.0		
			All information on non-existing spans will be deleted

Figure 6.2   *A screen dump from the Data Input spreadsheet for the continuous concrete beam*

shown in Figure 6.5). The results of this spreadsheet are then linked through to become the input for another spreadsheet calculating the quantities of main reinforcements, shear links and crack width etc. as shown in Figure 6.4. This minimizes the amount of input required and the scope for error in data transfers; for example, the results of a beam analysis can be carried through to beam design.

The use of graphics in spreadsheets can be manipulated without the need for any complicated programming knowledge. As the user changes certain input values, the graphics change immediately, showing both the revision and its effect.

4   Conclusion

The advantage of spreadsheets is that changing one cell changes all dependent cells. They are particularly suited to repetitive calculations and sensitivity analyzes. Engineers gain experience by doing their own 'what-if' studies. In trying out different options, the engineer can get a better feel for the effect that individual changes might have on the overall design. Spreadsheets are therefore a powerful tool for optimization of design and preparation of design calculations (Davies 1995, Orvis 1996, Liengme 2000).

Spreadsheets can be developed for particular solutions for which no commercial software exists. However, for commercial applications, any development and modification of spreadsheets should only be undertaken

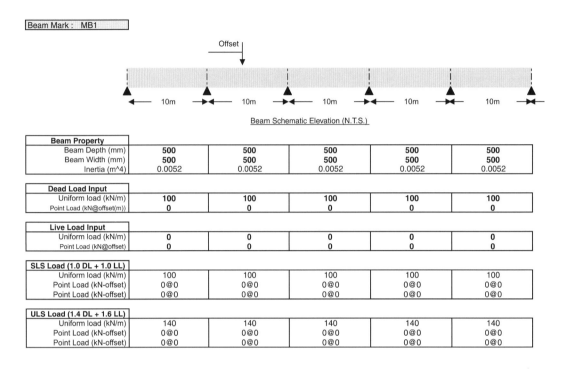

Beam Property					
Beam Depth (mm)	500	500	500	500	500
Beam Width (mm)	500	500	500	500	500
Inertia (m^4)	0.0052	0.0052	0.0052	0.0052	0.0052

Dead Load Input					
Uniform load (kN/m)	100	100	100	100	100
Point Load (kN@offset(m))	0	0	0	0	0

Live Load Input					
Uniform load (kN/m)	0	0	0	0	0
Point Load (kN@offset)	0	0	0	0	0

SLS Load (1.0 DL + 1.0 LL)					
Uniform load (kN/m)	100	100	100	100	100
Point Load (kN-offset)	0@0	0@0	0@0	0@0	0@0
Point Load (kN-offset)	0@0	0@0	0@0	0@0	0@0

ULS Load (1.4 DL + 1.6 LL)					
Uniform load (kN/m)	140	140	140	140	140
Point Load (kN-offset)	0@0	0@0	0@0	0@0	0@0
Point Load (kN-offset)	0@0	0@0	0@0	0@0	0@0

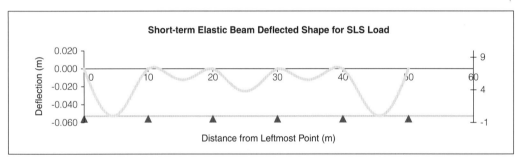

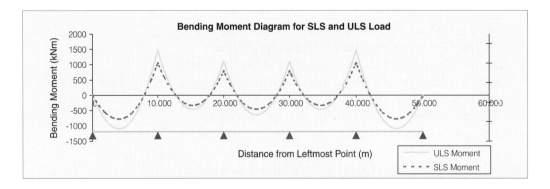

Figure 6.3   A screen dumps from the spreadsheet for the input and output of information for continuous beam analysis

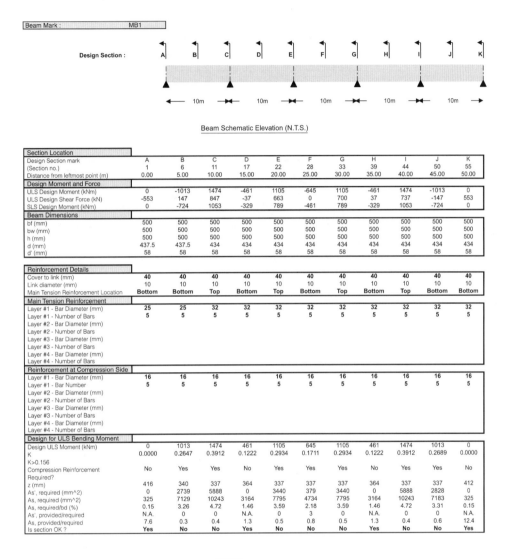

Beam Mark :	MB1

Design Section :   A | B | C | D | E | F | G | H | I | J | K

10m 10m 10m 10m 10m

Beam Schematic Elevation (N.T.S.)

Section Location											
Design Section mark	A	B	C	D	E	F	G	H	I	J	K
(Section no.)	1	6	11	17	22	28	33	39	44	50	55
Distance from leftmost point (m)	0.00	5.00	10.00	15.00	20.00	25.00	30.00	35.00	40.00	45.00	50.00
Design Moment and Force											
ULS Design Moment (kNm)	0	-1013	1474	-461	1105	-645	1105	-461	1474	-1013	0
ULS Design Shear Force (kN)	-553	147	847	-37	663	0	700	37	737	-147	553
SLS Design Moment (kNm)	0	-724	1053	-329	789	-461	789	-329	1053	-724	0
Beam Dimensions											
bf (mm)	500	500	500	500	500	500	500	500	500	500	500
bw (mm)	500	500	500	500	500	500	500	500	500	500	500
h (mm)	500	500	500	500	500	500	500	500	500	500	500
d (mm)	437.5	437.5	434	434	434	434	434	434	434	434	434
d' (mm)	58	58	58	58	58	58	58	58	58	58	58

Reinforcement Details											
Cover to link (mm)	40	40	40	40	40	40	40	40	40	40	40
Link diameter (mm)	10	10	10	10	10	10	10	10	10	10	10
Main Tension Reinforcement Location	**Bottom**	**Bottom**	**Top**	**Bottom**	**Top**	**Bottom**	**Top**	**Bottom**	**Top**	**Bottom**	**Bottom**
Main Tension Reinforcement											
Layer #1 - Bar Diameter (mm)	25	25	32	32	32	32	32	32	32	32	32
Layer #1 - Number of Bars	5	5	5	5	5	5	5	5	5	5	5
Layer #2 - Bar Diameter (mm)											
Layer #2 - Number of Bars											
Layer #3 - Bar Diameter (mm)											
Layer #3 - Number of Bars											
Layer #4 - Bar Diameter (mm)											
Layer #4 - Number of Bars											
Reinforcement at Compression Side											
Layer #1 - Bar Diameter (mm)	16	16	16	16	16	16	16	16	16	16	16
Layer #1 - Bar Number	5	5	5	5	5	5	5	5	5	5	5
Layer #2 - Bar Diameter (mm)											
Layer #2 - Number of Bars											
Layer #3 - Bar Diameter (mm)											
Layer #3 - Number of Bars											
Layer #4 - Bar Diameter (mm)											
Layer #4 - Number of Bars											

Design for ULS Bending Moment											
Design ULS Moment (kNm)	0	1013	1474	461	1105	645	1105	461	1474	1013	0
K	0.0000	0.2647	0.3912	0.1222	0.2934	0.1711	0.2934	0.1222	0.3912	0.2689	0.0000
K>0.156 Compression Reinforcement Required?	No	Yes	Yes	No	Yes	Yes	Yes	No	Yes	Yes	No
z (mm)	416	340	337	364	337	337	337	364	337	337	412
As', required (mm^2)	0	2739	5888	0	3440	379	3440	0	5888	2828	0
As, required (mm^2)	325	7129	10243	3164	7795	4734	7795	3164	10243	7183	325
As, required/bd (%)	0.15	3.26	4.72	1.46	3.59	2.18	3.59	1.46	4.72	3.31	0.15
As', provided/required	N.A.	0	0	N.A.	0	3	0	N.A.	0	0	N.A.
As, provided/required	7.6	0.3	0.4	1.3	0.5	0.8	0.5	1.3	0.4	0.6	12.4
Is section OK ?	**Yes**	**No**	**No**	**Yes**	**No**	**No**	**No**	**Yes**	**No**	**No**	**Yes**

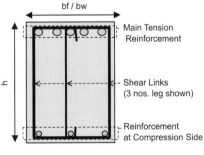

bf / bw

h

— Main Tension Reinforcement

— Shear Links (3 nos. leg shown)

— Reinforcement at Compression Side

Typical Beam Section (N.T.S.)
(For Section with Main Tension Bars at Top)

Figure 6.4   A screen dump from the spreadsheet for the input and output of information for reinforced concrete design

	1	2	3	4	5	6	7	8	9	10	11	12
						System Matrix K						
1	1500.0	7500.0	-1500.0	7500.0	0.0	0.0	0.0	0.0	0.0	0.0	0.0	0.0
2	7500.0	50000.0	-7500.0	25000.0	0.0	0.0	0.0	0.0	0.0	0.0	0.0	0.0
3	-1500.0	-7500.0	3000.0	0.0	-1500.0	7500.0	0.0	0.0	0.0	0.0	0.0	0.0
4	7500.0	25000.0	0.0	100000.0	-7500.0	25000.0	0.0	0.0	0.0	0.0	0.0	0.0
5	0.0	0.0	-1500.0	-7500.0	3000.0	0.0	-1500.0	7500.0	0.0	0.0	0.0	0.0
6	0.0	0.0	7500.0	25000.0	0.0	100000.0	-7500.0	25000.0	0.0	0.0	0.0	0.0
7	0.0	0.0	0.0	0.0	-1500.0	-7500.0	3000.0	0.0	-1500.0	7500.0	0.0	0.0
8	0.0	0.0	0.0	0.0	7500.0	25000.0	0.0	100000.0	-7500.0	25000.0	0.0	0.0
9	0.0	0.0	0.0	0.0	0.0	0.0	-1500.0	-7500.0	3000.0	0.0	-1500.0	7500.0
10	0.0	0.0	0.0	0.0	0.0	0.0	7500.0	25000.0	0.0	100000.0	-7500.0	25000.0
11	0.0	0.0	0.0	0.0	0.0	0.0	0.0	0.0	-1500.0	-7500.0	1500.0	-7500.0
12	0.0	0.0	0.0	0.0	0.0	0.0	0.0	0.0	7500.0	25000.0	-7500.0	50000.0

	1	2	3	4	5	6	7	8	9	10	11	12
					Inverse of reduced System Matrix K							
1	50000	25000	0	0	0	0	0	0	0	0	0	0
2	25000	100000	25000	0	0	0	0	0	0	0	0	0
3	0	25000	100000	25000	0	0	0	0	0	0	0	0
4	0	0	25000	100000	25000	0	0	0	0	0	0	0
5	0	0	0	25000	100000	25000	0	0	0	0	0	0
6	0	0	0	0	25000	50000	0	0	0	0	0	0
7	0	0	0	0	0	0	1	0	0	0	0	0
8	0	0	0	0	0	0	0	1	0	0	0	0
9	0	0	0	0	0	0	0	0	1	0	0	0
10	0	0	0	0	0	0	0	0	0	1	0	0
11	0	0	0	0	0	0	0	0	0	0	1	0
12	0	0	0	0	0	0	0	0	0	0	0	1

Figure 6.5   A screen dump from a spreadsheet for inverting the global stiffness matrix

by experienced engineers who have a good understanding of the problems and pitfalls of design and spreadsheets.

There are questions concerning the quality assurance of spreadsheets, particularly in commercial applications. Although the contents of spreadsheets can be easily altered or modified to suit one's needs, this can be a dilemma in terms of quality assurance. To avoid the inadvertent corruption of formulae and reduce the scope for errors during use, cells can be protected. In addition, spreadsheets should be treated as controlled documents under a QA system. In any circumstances, as with all computer software, one should always bear in mind that spreadsheets are not foolproof and the users should always check results against approximate answers and exercise professional judgement in assessing the results.

References

1. Cheung, Y. K., Lo, S. H. and Leung, A. Y. T. 1996. *Finite Element Implementation*. Oxford: Blackwell Science.

2. Davies, S. R. 1995. *Spreadsheets in Structural Design*. Wiley, New York: Longman Scientific and Technical.

3. Kong, F. K. and Evans, R. H. 1987. *Reinforced and Prestressed Concrete*. London: Chapman and Hall.

4. Liengme, Bernard V. 2000. *A Guide to Microsoft Excel for Scientists and Engineers*. Oxford: Butterworth-Heinemann.

5. Look, Burt. 1994. *Spreadsheet Geomechanics: An introduction*. Rotterdam: AA Balkema.

6. Macleod, I. A. 1988. *Guidelines for checking computer analysis of building structures*. London: Construction Industry Research and Information Association, Westminister.

7. _____. 1990. *Analytical Modeling of Structural Systems*. West Sussex: Ellis Horwood.

8. Meyer, Christian. 1990. *Finite Element Idealization*. New York: American Society of Civil Engineers.

9. Orvis, William J. 1996. *Excel for Scientists and Engineers*. California: SYBEX.

10. Zienkiewicz, O. C. and Taylor, R. L. 1989. *The Finite Element Method (Volume 1)*. Singapore: McGraw-Hill International Editions.

Chapter 7

Advanced Simulation Tools for Building Services Design

Tin-tai CHOW and Apple Lok Shun CHAN

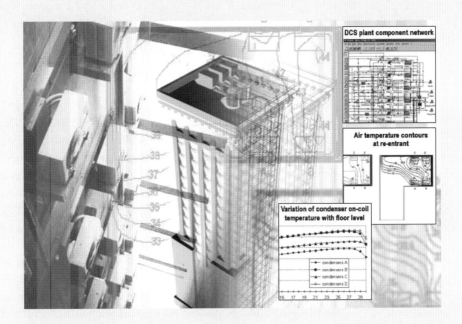

Worldwide energy and environmental concerns have prompted the local construction industry to pay closer attention to building sustainability and the Hong Kong SAR Government to put forward new codes and regulations in governing building performance. These new developments in the industry progressively raise the demands on the use of advanced simulation tools for comprehensive performance analyzes. This chapter introduces the state-of-the-art computer simulation tools that are widely applicable in building services designs, such as in building thermal analysis, airflow prediction, and visual/acoustical comfort evaluation. Case studies are used to illustrate how they work in actual practices. Their present needs and future developments are also addressed.

7

Advanced Simulation Tools for Building Services Design

1 Introduction

1.1 Modernization of building design

Building design has undergone huge changes in Hong Kong; new buildings are bigger and taller and they penetrate deeper underground. In commercial buildings, curtain walling, atriums and information technology provisions have become an integral part of modern design. Developers like to build a mix of building types, such as commercial centres, office towers, hotels, and residential towers, within the same development. In the last 20 years or so, the global energy crisis and environmental issues have made conservation of energy a key factor in architectural design. Since the public became aware of sick building syndrome in the 1980s, indoor air quality has also become an important design factor. Obviously, a building featuring contemporary aesthetics is far from modern if the quality of the building services do not match. Taller and larger buildings present an increasing challenge in the provision of comfortable yet cost-effective artificial environment and impose increased pressure on the means of evacuation and fire-fighting. Thus, building design in terms of human comfort, safety, health, and energy use must be precisely evaluated.

In response to these developments, the Hong Kong SAR Government has progressively put forward new codes of practice and regulations on energy and environmental performance of buildings and services installations in recent years. There are increased efforts to introduce innovative building design and new concepts in building services installations. There is also a trend towards performance-based system designs, such as smoke ventilation systems and OTTV (Overall Thermal

Transfer Value) determination. The industry is calling for the construction of buildings that are equipped with "hi-tech" intelligent features that meet specific environmental assessment standards. There should be wider use of renewable energy and more promotion of building sustainability. All these have created a demand for the use of advanced predictive tools for comprehensive performance analyzes.

1.2 Advanced predictive tools

Building performance analyzes rely on the use of physical or laboratory studies and/or numerical simulation in one way or another. Physical model studies with wind tunnels, environmental chambers and artificial sky domes are popular in the context of both practical design and academic research. Wind tunnels are very effective in the study of wind effects, such as its impact on building structure, wind pressure at openings in relation to ventilation design, and the dispersion of pollutants over an urban district. Environmental chambers are important in the study of indoor air movement and human comfort. Artificial sky domes allow the assessment of the effect of overshadowing and the control of sunlight penetration. The limitations of these laboratory approaches are that they are relatively expensive and the required time-span can be lengthy. Laboratory results are also subject to experimental errors related to physical modelling and measurement techniques. Computer analysis, on the other hand, is more economical and takes less time, making it preferable when prompt decisions have to be made.

Computer technology has achieved major advances since the first computer was made in the 1950s. The exponential increase in computing speed and storage plus the corresponding drop in price have made it possible to develop advanced graphical and numerical tools to handle complex analytical tasks. This has led to the widespread use of simulation software in the construction industry. Many public domain and commercial software packages with sophisticated user interfaces are now available. Extensive model validation and comparative studies have been conducted with some tests performed by reputable international organizations with participating software firms from around the world. The fast-growing use of simulation software has gradually won the trust of the construction industry in simulation technology. Practitioners today generally accept that modelling-and-simulation is a practical approach towards assessment of building performance.

Design analysis supported by computer simulation involves the "creation" of a behavioural model of a building in a given project stage. The simulation task involves executing this model on a digital computer and analyzing its observable states, which are made up of the post-processed outputs of the simulation runs. Behavioural models are developed by reducing real world physical entities and phenomena to an idealized form and at a desired level of abstraction. From this abstraction, a mathematical model is constructed by applying the conservation laws of physics. The models derived range in type from empirical to fundamental.

1.3   Areas of application in building services

Electrical and mechanical systems such as heating, ventilation and air-conditioning (HVAC), fire services, plumbing and drainage, electrical services and transportation systems form the core of building services engineering. Computer simulation tools are widely applied in the decision-making process in association with building services design. Typical areas of application are outlined below.

Building thermal analysis

Heat and mass transfer processes in air-conditioning systems remove unwanted heat and moisture from the indoor space. Computer simulation techniques can provide accurate estimation of hourly heat gains and cooling loads, allowing engineers to generate an optimum design for the HVAC system with proper equipment of the right size. At the same time, architects can conduct parametric studies on different building features for screening energy-efficient design options. As a matter of fact, building envelope and environmental services provisions may be regarded as an integrated object for thermal performance evaluation.

The "Energy Standard for Buildings except New Low-rise Residential Buildings" published by the American Society of Heating, Refrigerating and Air-conditioning Engineers, Inc (ASHRAE) provides a performance-based Building Energy Cost Budget Method for building services and building envelope design. Trade-offs in the energy performance of various building services systems and the building envelope are allowed so that innovative building services and building designs become technically feasible. Computer simulation can be conducted to determine the annual energy cost of a newly designed building so that its compliance with the energy performance-based code can be assessed.

Another type of building thermal analysis is used to simulate the transient performance of thermal energy systems such as refrigeration plant or the HVAC system. Here, component-based simulation software is used to evaluate the optimum system configuration, control strategy and sequencing of the central air-conditioning plant.

Air flow prediction

Air temperature, humidity, velocity and pollutant concentrations are the key parameters for assessing indoor thermal comfort and air quality. Recommended design values of these parameters are readily available from design guides and handbooks, so the HVAC system can be selected and developed accordingly. Even when these recommended values are generally satisfied in bulk in an indoor space, local thermal discomfort and poor air quality may still exist in specific locations. These problems may be caused by improper design solutions in a number of areas, such as terminal unit location, sensor location, supply air temperature, and ventilating airflow rate.

Patterns of indoor airflow and pollutant distribution can be predicted by solving a set of partial differential equations for conservation of mass, momentum, energy and species concentration. Computational Fluid Dynamics (CFD) is a numerical technique applied for this purpose with some given boundary conditions. The results can be used to assess the appropriateness of, among others, the air distribution system and building material selection. A major advantage of using computer analysis over wind tunnel on fluid flow is that it provides results for the entire flow field, rather than just at points on a physical model where pressure gauges are placed. It is particularly useful for applications such as sensitivity analysis of unconventional buildings where there is little prior experience, and for cases where the complete information about the whole flow field is important.

A clear understanding of the behaviour of fire and smoke propagation is crucial for the provision of escape routes in buildings. CFD simulated results can provide useful information for fire services engineers designing and evaluating the performance of various fire services systems such as fire detection, sprinkler and smoke extraction systems.

Visual/acoustic comfort evaluation

General lighting criteria for visual comfort includes appropriate task illuminance, relative brightness (contrast) and avoidance of glare. Artificial lighting provides reliable illumination for indoor space regardless of the outdoor daylight condition. In some buildings, daylight integrated with artificial lighting is used to reduce electricity consumption. However, natural lighting in an indoor space may lead to glare and increased air-conditioning load. Accordingly, detailed studies are often conducted in the design stage to investigate the degree of visual comfort and the impact on electricity consumption in buildings. Lighting performance simulation software can fulfil the need to predict interior illuminance in complex building spaces under a variety of lighting conditions.

Computer simulation can also be used to study noise interference and sound quality in facilities such as conference rooms, concert halls and TV studios where optimum reverberation time and acoustic performance in terms of echo, sound shadow and long delayed reflection are of crucial importance. Machine plant rooms, in particular the emergency diesel generator room, should be acoustically insulated. In the past, time-consuming and expensive scale models were used to evaluate the acoustic design of buildings. Nowadays, with advanced computing technology, computer simulation programs using ray-tracing technique have enabled design engineers to predict and assess the acoustic behaviour of any enclosed or open space with various building materials and under various operating conditions.

2   Building Performance Simulation Software

2.1   Building thermal analysis software

Thermal analysis programs for buildings have been available for many years. They vary from simple methods that can be performed at the initial design stage using spreadsheet through to sophisticated dynamic thermal simulations. There are two major types of dynamic simulation software for building thermal analysis, namely building energy simulation programs and building thermal systems simulation programs.

Building energy simulation programs

Building thermal load varies with time due to natural changes both indoor and outdoor, as well as the heat capacity of the building structure that gives rise to thermal effects on a time-dependent manner. Building elements such as external walls, roofs and floor slabs (that work as "capacitors") play an important role in delaying the effects of heat flow. It is necessary to have precise estimates of hourly heat gains through the building envelope as well as cooling loads on an unsteady state basis. Heat Balance (HB) equations and Radiant Time Series (RTS) are two commonly used approaches to making precise estimates (ASHARE 2001a).

There are a number of building energy simulation software available in the market, e.g. DOE-2, BLAST, and EnergyPlus. DOE-2 employs room weighting factors for the calculation of thermal loads and room air temperatures, while BLAST calculates hourly space cooling loads with the room heat balance approach. These two programs have been widely adopted for studying the thermal performance of different building envelope designs. Parametric studies can be conducted with different combinations of window-to-wall (WWR) ratios, façade orientations, the shading coefficients of glazing, aspect ratios, thermal transmittances of opaque walls and roofs, etc. These programs have sub-programs containing a database of built-in air-side systems and central air-conditioning equipment for users to select and incorporate into the building. With the input data, the hourly electric and hot/chilled water demand of the air-handling system, the energy consumption of the central plant as well as the whole building's energy consumption can be evaluated. For new buildings, these results can be used to assess compliance to performance-based energy codes while the energy target (in terms of kWh/m^2) for existing buildings can be established for energy monitoring. Any substantial deviation of the measured building energy consumption from the energy target will send an alert signal to the building owner or operator, triggering action to check whether the deviation in energy consumption is abnormal or acceptable due to, for instance, a change in building operation schedule or outdoor weather conditions.

In 2001, a new-generation building energy simulation program called EnergyPlus was introduced. It combines the best capabilities and features of DOE-2 and BLAST with new features, one of which is a newly developed public domain simulation software with a modular structure that facilitates the adding of features and links to other programs such as those used for daylight and HVAC simulations. The inclusion of fluid/air loops and user-configurable HVAC systems makes EnergyPlus more or less a building thermal systems simulation program, as described below.

Building thermal systems simulation programs

These programs, also known as system component-based simulation, are provided with a system component library. They can fulfil the tasks performed by the building energy simulation programs mentioned above. In addition, they can provide HVAC system-only simulation or integrated building-and-system simulation. The user can link together the various components to form a thermal system of specific desirable topology. The simulation engine then calls the system components and iterates the simulation based on the input files until the system equations are solved. TRNSYS, ESP-r and HVACSIM+ programs belong to this category. To suit individual simulation tasks, users are allowed to construct their own system component models such as air-handling units, water pumps, chillers, and storage tanks, with specific physical properties including dimensions and heat transfer coefficients.

The modular feature and capability of this type of simulation program allow users to conduct simulations for solar thermal systems, fuel cells and modern renewable energy systems such as photovoltaic and wind power systems. In building services engineering, various HVAC components and systems can be modelled and simulated. For example, various heat rejection systems like air-cooled condenser and cooling tower using potable water or seawater can be modelled and their energy performance as well as running cost can be evaluated. With the aid of lifecycle costing, the best option can then be chosen for retrofit of an existing HVAC system.

The design of the routing of air/water distribution loops in an HVAC system is mainly based on the criteria of space constraint, minimum pressure loss and the design engineers' own experience and judgement. With system component-based simulation software, the distribution loops can be modelled and the design optimized, taking into account the conduit dimensions, pressure drop, heat gain/loss as well as the dynamic interaction with the central plant under designed control.

2.2   Computational Fluid Dynamics(CFD) software

CFD is a sophisticated flow analysis technique extensively used both as a design tool and for research purposes. It not only predicts fluid flow behaviour, but also the transfer of heat, mass, phase change, chemical reaction, and stress and deformation of solid structures under the fluid's pressure. First a computational model is developed through which a target system or device under investigation is represented. The fluid flow physics is then applied to this virtual prototype. The output is a prediction of the fluid dynamics.

CFD gives a means of visualizing and enhancing understanding of the designs. The term usually refers to numerical solutions to the Reynolds-averaged Navier-Stokes equations, which means that averaged velocities and temperature, among others, are predicted for turbulent flow. The analysis is able to show one part of the system under investigation, or phenomena happening within the system, that otherwise would not be visible through any other means. CFD facilitates better and faster design by testing different options until an optimal result is achieved. Most commercial CFD software can be run on the new generation of desktop computers. General-purpose commercial CFD packages are available on the market, such as CFX, PHOENIX and FLUENT. Almost all building simulation methodologies use finite volume techniques for the CFD calculation, though finite element is the alternative choice and is more suitable for complicated geometry. There are three main groups of airflow problems in buildings: calculation of the external flow around buildings, calculation of the bulk internal flow, and the transient flow simulation. The two-equation k–ε model with Boussinesq approximation is by far the most popular technique. Steady incompressible flow is often assumed, taking into account turbulence and buoyancy, but in principle any type of unsteady (transient) flow can be treated. The technique has an inherent weakness in that it relies on empirical equations to simulate turbulence. It then becomes very important to choose the correct boundary conditions and the correct grid discretisation scheme (that can be either uniform or non-uniform), in particular at the regions close to the walls. The user is expected to have a good feel of the outcome of the flow pattern that could result.

The application of CFD to non-spreading fires in buildings requires additional conservation equations to those required in the above studies. More complex models employ equations describing the chemical reactions during combustion and include radiation in the energy equation.

2.3   Lighting performance software

Lighting performance simulation software provides an effective means of predicting illumination conditions for complex building space with electric lighting and/or daylight. Well-known lighting performance simulation programs include SUPERLITE and RADIANCE. The former is a classic daylight simulation program that uses the flux transfer method as its simulation approach. It is capable of calculating daylight factors and daylight levels on any given plane for detailed room geometries, standard daylighting techniques and shading from external and internal obstructions. RADIANCE uses the ray-tracing technique to calculate the luminance in a furnished space of any shape. It can also create a three-dimensional (3D) and colour-rendered visual representation of space and furniture with shading and calculated luminosity values.

A new and integrated lighting simulation tool called ADELINE allows the conversion of the CAD drawing of a space into its input file and links to the above-mentioned lighting performance simulation programs. It predicts and presents 3D displays of various lighting scenarios in colour, with complex graphic representations, to facilitate analysis of luminance distribution, glare sources and visual comfort.

The output results of lighting performance simulation programs can be used as input data for building thermal performance programs. This helps the study of the energy impact of daylight on buildings and HVAC systems. For example, a room which utilizes daylight can be divided into a number of zones and the simulated lighting level for each zone is compared to a design value, say 500 lux. Where a zone is over-lit, stepped lighting or continuous dimming control can be applied and functions could be generated to describe the relationship between light output and electrical input into the lighting system. Therefore, 8,760 hourly fractions of electric lighting lit during daylight hours in all the zones are calculated and transferred to the building thermal performance program as an input file. The data enables the program to determine the reduction in electricity consumption by electric lighting and the corresponding increase in space cooling load. The electricity consumption of the HVAC system can thus be calculated. With these simulated results, design engineers can have a clear picture of where daylight should be utilized in a building for achieving energy conservation.

Three more examples of the application of advanced simulation software in environmental services design in Hong Kong are elaborated in the following section.

3   Application Examples

3.1   Comparative studies in building envelope performance

In an energy-efficient building, heat gain and heat loss through the building envelope is minimized. There are two main categories of energy codes: prescriptive and performance-based. In Hong Kong, the Code of Practice for Overall Thermal Transfer Value (OTTV) in Buildings is one typical example of performance-based energy code. It specifies a maximum OTTV value of 30 W/m^2 for a building tower and 70 W/m^2 for a podium. This approach allows building designers the freedom to vary building envelope features in order to meet the specific design objectives and comply with the OTTV requirement. Key design parameters involve building orientation, insulation, window glazing types, window-to-wall ratio (WWR), building aspect ratio, etc. Building energy simulation program is a useful tool in the sensitivity study of building envelope performance. With additional input data from the air-conditioning system and central plant equipment, monthly and annual electricity consumption of the whole building under typical weather conditions can be evaluated. The results are of great use in establishing the energy target for a building and in serving as a guide for building operators and maintenance engineers monitoring the monthly and annual energy consumption of a building. Any abnormal deviation from the target can therefore be investigated and appropriate action taken.

Take a 40-storey generic curtain-wall commercial building in Hong Kong as an example to explore the relative thermal performance of the building envelope. In the base case, 0.5 is used for the shading coefficient (SC) of window glazing, 40% for WWR, with no thermal insulation installed in the external wall. Table 7.1 lists the base case design parameters (**shaded**), together with other parameters for comparative studies. The public domain software EnergyPlus was used in the study to evaluate the heating and cooling loads of this generic building with various design parameters in separate simulation runs. The simulated annual heating and cooling loads are shown in Figures 7.1 and 7.2 respectively, for ready comparison with the base case design. As seen from Figure 7.1, the increase in heating load with glazing of SC equal to 0.1 and 0.3 are 32% and 14% respectively, while there is a 11% reduction in heating load for glazing with SC = 0.7. This is mainly due to the reduction in solar heat gain through window with glazing of lower SC in winter. In cases with larger WWR, despite the increase in solar heat gain through larger window glazing area, the increase in heating loads is 17% and 34% respectively because the larger window area has lowered the thermal resistance of the whole exterior wall, resulting in more heat loss by convection and conduction. Thermal insulation is crucial in the reduction of heat loss through opaque walls. The decrease in heating loads with 25mm and 50mm insulation is 30% and 42% respectively.

Figure 7.2 shows the increase in cooling load with increases in SC and WWR. A higher SC and larger window area (both having significant effect on the increase in solar heat gain) were found to contribute much to the increase in the cooling load of the building. The increase is 8% (for SC = 0.7) and 13% (for WWR = 80%) respectively. For thermal insulation inside opaque wall, it was found that the reduction in cooling load is not significant. There is only a 2% and 3% decrease in cooling loads with 25mm and 50mm insulation respectively.

3.2   District cooling system assessment

An integrated building and plant simulation can be used to evaluate the huge investment in chiller plant and distribution network of a district cooling system (DCS). DCS involves the use of centralized chilled water generating plant and pipe distribution network to deliver cooling energy to different buildings within a district. Chilled water is produced in a central plant with groups of chillers and pumping station and is circulated to the buildings through underground supply and return pipes. DCS has economic advantages because the total or peak installed cooling capacity of a DCS plant is smaller than the sum of conventional individual plants at the consumer buildings. The centralized plant also benefits from the

	Parameters	
Shading Coefficient	Window-to-Wall Ratio	Insulation in Opaque Wall
0.1	20%	**No insulation**
0.3	**40%**	25mm insulation
0.5	60%	50mm insulation
0.7	80%	—

Table 7.1   *Key parameters of building envelope in the comparative studies*

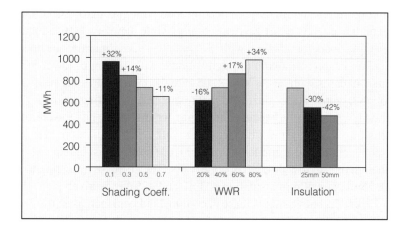

Figure 7.1   *Annual heating loads of the generic building with different shading coefficients, WWR and insulation*

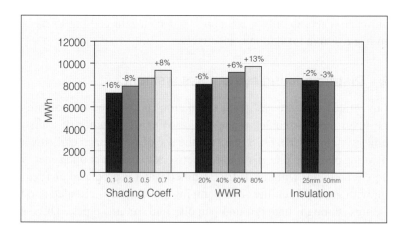

Figure 7.2   *Annual cooling loads of the generic building with different shading coefficients, WWR and insulation*

higher efficiency associated with larger chillers. Plant spaces in buildings can also be released for other purposes. Because of population density, DCS technology has a good potential for application in reclaimed areas and sites planned for a complete redevelopment, such as the South East Kowloon Region. The chillers can be of the direct seawater-cooled type if the central plant is located close to the sea front.

In DCS design, the key considerations are the required cooling capacity of the different phases of a project, the location of the seawater pump station, the location of the central plant, the routes of the piping network, and the adoption of design options such as thermal storage, co-generation and tri-generation schemes. These design considerations can be readily investigated by means of building thermal analysis (Chow et al. 2001). Figure 7.3 shows the flow chart in the design process. There are two main steps, namely thermal load estimation and energy use analysis. A database of the space thermal loads of the targeted DCS service purchasers in a district is first developed. A building categorization scheme is implemented so that the identified buildings are divided into groups according to the nature of their operation and functional use. The usable floor area of each building category can then be established and typical buildings in each category can be modelled. The peak thermal loads in design months and the year-round cooling load pattern of the group can then be estimated using building energy simulation software. From the results, normalized design load profiles and year-round cooling/heating load profiles of each category in W/m^2 can be determined. For a development with buildings in various categories, the hourly cooling load profile of the design month can be determined by summing up the product (multiplication) of the normalized cooling load by the gross floor area of each category. Then the maximum hourly cooling load can be used to determine the required capacity of the DCS plant. A similar approach can be applied for derivation of year-round cooling and heating loads profiles. These simulated results are then used for evaluating the different DCS scheme options and the required capacities of individual plant components.

The location of the DCS central plant must be strategically investigated in order to minimize pipe lengths and hence the pumping power. Moreover, a range of inter-related technologies such as thermal storage, waste heat recovery, tri-generation, and chilled water looped circuit pipework can be incorporated into the DCS model for specific application. Building thermal system simulation software enables design engineers to model the DCS with these design options. Based on the simulation results, the hourly electricity demand and annual energy consumption of the chiller plant and hence the optimum DCS design for a specific application can be determined.

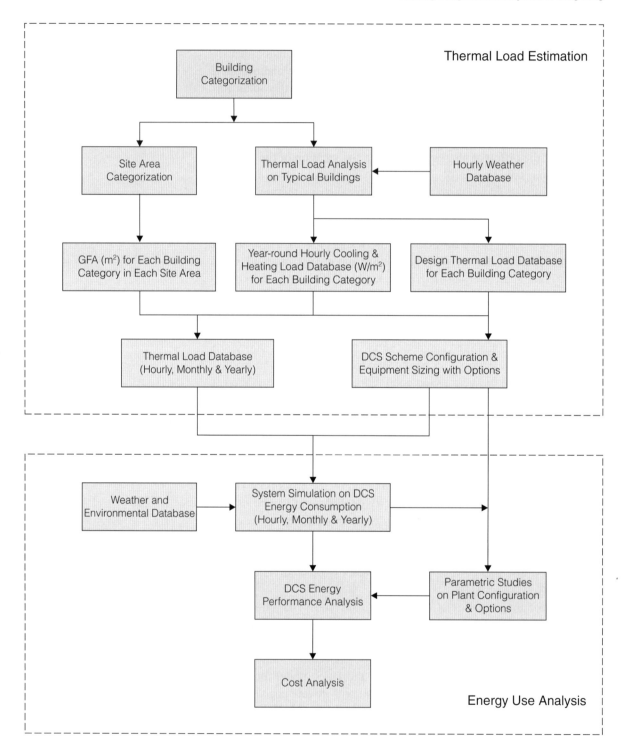

Figure 7.3   Flow chart of DCS design process

In our plant simulation example, the hourly electricity consumption of the major equipment in the DCS plant, including chillers, production and distribution water pumps and seawater pumps, are estimated by using TRNSYS. Each individual component of the DCS plant is modelled and then interconnected to form the circuit (see Figure 7.4 for reference). With the hourly thermal loads in the district simulated by DOE-2, environmental data such as seawater temperature and the hourly system performance of the DCS plant can be evaluated. The same simulation procedure can be adopted for the entire period of the planned development. Figure 7.5 shows an example of the annual electricity consumption of a cooling-only DCS plant in different years of the hypothetical urban development. The output of the simulation, i.e. the year-round hourly data of the electricity consumption profiles, can be transferred to a cost model to analyze the annual operation costs under different tariff structure. The cost model can serve dual purposes: (i) as a financial modelling tool to address the cost-effectiveness of each of the DCS options, and (ii) as a tool to analyze the bids received from potential operators at the tendering stage of the implementation process.

3.3   Airflow in building re-entrants

Residential building projects in Hong Kong typically comprise a number of high-rise residential towers standing on a common podium. The pattern of tower arrangements generally varies depending on the prevailing climate and topography. Figure 7.6 shows an example with eleven residential towers above a common podium garden. Developments based on a "cruciform" plan for individual towers is essentially a product of the Building (Planning) Regulations. Public space including lift lobbies and corridors is usually squeezed to a minimum in the building core. With its four wings radiating out from the core, a residential tower consists usually of eight apartments on a single floor, with two in each wing. A narrow but deep re-entrant exists between the neighbouring flats of each wing. The windows of kitchens and bathrooms that are open to a re-entrant are meant to facilitate natural lighting and ventilation. Because of restrictive air movement at the re-entrant, oily kitchen exhaust is frequently trapped within the re-entrant in adverse weather. This results in extensive re-circulation back into the kitchens via opened windows. In new residential buildings, a growing trend is to provide split-type air-conditioners for living/dining rooms and bedrooms. From an aesthetic point of view, the recessed building re-entrants are particularly well suited for accommodating the outdoor condensing units. However, the thermal energy dissipated from the stacks of the condensing units tends to produce a buoyant plume of airflow. This results, on the one hand, in an elevated

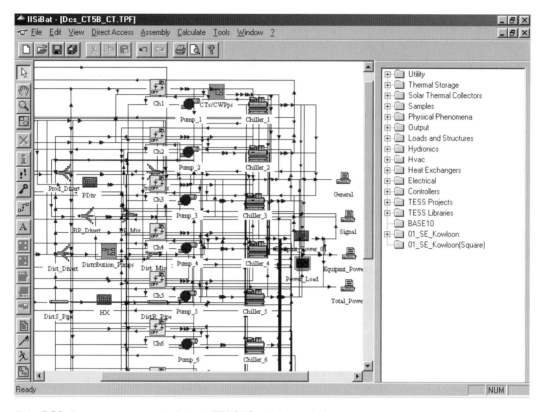

Figure 7.4   DCS plant component network under TRNSYS simulation platform

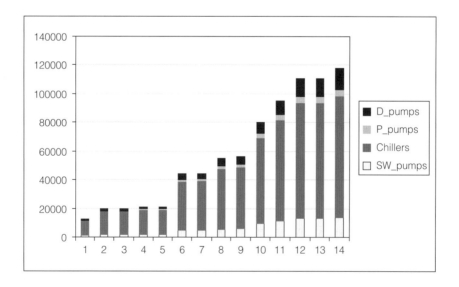

Figure 7.5   Annual DCS electricity consumption for a new urban development

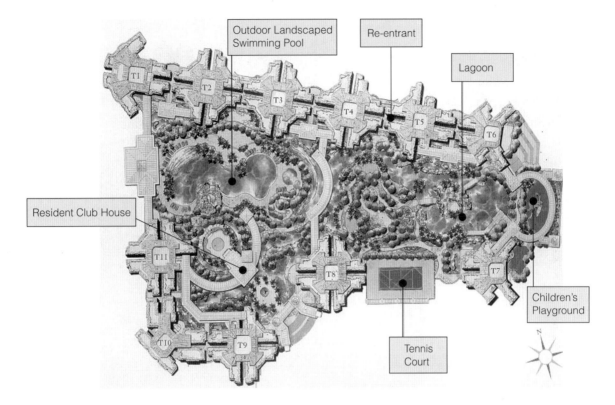

Figure 7.6   Plan view of a residential project in Hong Kong

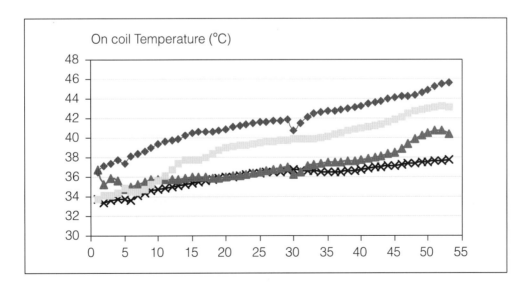

Figure 7.7   On-coil temperature of six condensing units vs floor level at a building re-entrant

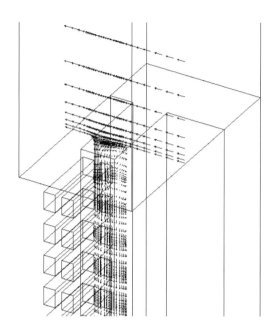

Figure 7.8   Velocity distribution at vertical plane 2m from re-entrant end-wall under steady wind and with condenser-off

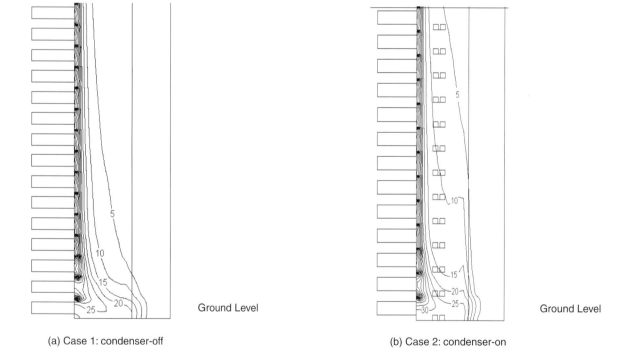

(a) Case 1: condenser-off   (b) Case 2: condenser-on

Figure 7.9   Percentage concentration of kitchen exhaust contaminants under steady wind situation

temperature in the re-entrant and, on the other hand, in a greater amount of air exchange that consequently reduces the re-circulation of kitchen exhaust.

As far as the operation of split-type air-conditioning units is concerned, an elevated temperature environment in the re-entrant significantly affects the condenser performance on the upper floors. The result could be an overall degradation of the capacity and efficiency of the air-conditioners of the entire building. As most condensing units are unlikely to function properly with an on-coil temperature (incoming cooling air temperature) above 45°C, the drawback is therefore not only energy wastage but also equipment malfunction. Accordingly, an HVAC engineer has to assess whether certain arrangements of the condensing units at a re-entrant will work properly even under the expected worst operating condition. This can be achieved by using CFD to study the re-entrant airflow and temperature distribution. Simplifications and assumptions have to be made to develop a steady-state computational model to facilitate the analysis (Chow, Lin and Liu 2002a). Figure 7.7 shows a measure of the condenser on-coil temperature against floor level for six different condensing units in a re-entrant with 53 floor levels (hence 6 × 53 or 318 condensing units working together), assuming all air-conditioners operate at a 50% load in the absence of wind and the outdoor temperature is 33°C. The effect of wind on air-conditioner performance or dilution of kitchen exhaust can be predicted (Chow, Lin and Liu 2002b). Figure 7.8 shows the airflow pattern around the top part of the re-entrant in the presence of moderate wind and when the air-conditioners are turned off. Figure 7.9 shows the percentage concentration of kitchen exhaust contaminants under a steady wind, with 9(a) showing the "condenser-off" situation, and 9(b) the "condenser-on" situation. Hence, given different wind conditions, pollutant exhaust conditions and the level of condenser operation, a detailed assessment of air-conditioning unit performance and kitchen ventilation can be achieved.

4   Present Needs and Future Development of Simulation Tools

4.1   Industrial needs

The role of simulation tools in the design and engineering of buildings has been firmly established over the last two decades. Simulation is credited with speeding up the design process, increasing efficiency, and enabling the comparison of a broader range of design variants, leading to more optimal solutions. Future trends in the development of building simulation tools are driven by the need for better support of design decisions and better quality control over performance assessments. The progression from a simple predictive tool to a more comprehensive one is welcome by most users because of the extended functionality and accuracy. For advanced simulation tools, a moderate level of proficiency requires formal training as well as hands-on experience from weeks to months. A common hope among building professionals is to be able to work with user-friendly software that can give correct answers in a relatively short time.

In principle, all software and hardware tools should only require from the user full knowledge of the physical problem under consideration and how to use the tool. If an in-depth knowledge of the software's methodology and excessive interaction at a specialized level is required, then the methodology is not "ready" and improvements must be made to bring it to a level where a practising designer who understands the physics of the phenomenon and has reasonable knowledge of the methodology can use it. This may not happen in the near future. The efficacy of dynamic thermal simulation tools in the design office or consulting practice is dependent not only on the facilities offered by the tools and the rigour of the underlying calculations, but also on the skills of the user in abstracting the essence of the problem into a model, setting up simulations and interpreting their results. Along this line, it is important for a qualified user to be familiar with the thermal properties of building materials and plant equipment. One should understand the exact meanings of the parameters to be input into the program as well as the importance of time step selection. In the case of CFD tools, in spite of the great effort to develop user-friendly packages, their efficient use still requires an expert user. It is very important to use the appropriate boundary conditions as well as grid discretisation. How the users are able to acquire the skills mentioned above, and how simulation tools can be made more accessible are important areas of future software development.

4.2 Integrated modelling

Using a design tool that focuses on a single domain is likely to result in sub-optimum design solutions. It is more appropriate, where possible, to use an integrated simulation tool throughout the design process than to use a progression of tools, from simplified to detailed, and to ignore the many theoretical discontinuities and even contradicting assumptions. While the components of a thermal model — the building, the plant, the CFD domain, etc. — may be processed independently, it is better to subject them to an integrated assessment whereby the dynamic interactions are explicitly represented. One example is the application of a façade-integrated photovoltaic/hot water collector system; a brief circuit diagram is shown in Figure 7.10. A solar PV/hot water collector system is able to generate electricity and hot water simultaneously. When compared to a conventional solar collector, the overall thermal efficiency is improved because of the multi-production and the cooling effect of the PV module. On the other hand, savings in cooling energy can be considerable if the external walls are covered with hybrid solar collectors. A numerical study of the active and passive thermal performance of the façade-integrated solar collector system can help to estimate the annual availability of electricity and hot water and to predict the annual reduction of air-conditioning electricity consumption in the building. As the heat transfer at the collector glass surface relies on wind velocity, an accurate hourly performance analysis requires a model integrating building, plant, electric power flow and airflow.

Thermal simulation and CFD programs provide complementary information about the performance of buildings in a number of ways. By means of thermal simulation, space-averaged indoor environmental conditions, cooling and heating loads, coil loads, and energy consumption can be obtained on an hourly or sub-hourly basis for periods ranging from a design day to a reference year. CFD programs, on the other hand, make detailed predictions of thermal comfort and indoor air quality, including the distribution of air velocity, temperature, relative humidity and contaminant concentrations. The distribution can be used to determine indices such as the predicted mean vote (PMV), the percentage of people dissatisfied (PPD) due to draft, and ventilation effectiveness. With the information from both approaches, engineers can design environmental control systems for buildings that satisfy multiple criteria.

The aim of integrated modelling is therefore to preserve the integrity of building and plant systems by simultaneously processing all energy/mass transport paths at a level of detail commensurate with the objectives of the problem in hand and the uncertainties inherent in the describing data. The ESP-r program, for instance, is currently evolving in this direction (Clarke 2001). In this sense, a building and the associated systems should be regarded as being systemic, dynamic, non-linear and, above all, complex. The following is a description of an ideal comprehensive simulation tool.

4.3   Ideal comprehensive tool

Figure 7.11 shows the possible structure and main features of an ideal comprehensive building performance evaluation tool. The simulation

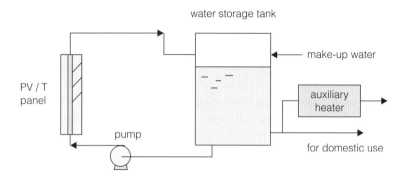

Figure 7.10   Water-heating Circuit in Photovoltaic/thermal (PV/T) Collector Scheme

program can incorporate conversions from CAD files which are used to create a building model of arbitrary complexity. This model can be imported from the "Project Manager" (intelligent user interface), where the form of problem and data entry can be interactive and knowledge-based. The Project Manager module gives access to support databases, simulation engine and core, which includes a number of performance assessment tools and a variety of third party applications for various purposes such as visualization and report generation. Its function is to co-ordinate problem definition, and give and receive the data model to/ from the simulation engine when the design hypothesis changes. Flexibility should be offered to allow the users more freedom to select or re-construct component models to suit various simulation tasks. The computation is processed in the simulation core, which also houses a number of inter-related sub-programs to undertake mechanical routine calculations. The systems of equations are solved numerically, the process of which can be direct or iterative, to produce the calculation results. Selective analysis or post-processing of these results, with due consideration for related factors such as human physiological responses, specified design standards and optimization goals, produces the output in suitable reporting formats.

Further extension of the software to include other building services areas like fire services, plumbing and drainage, or noise assessment is possible, depending on whether the integration will benefit the accuracy of the overall performance evaluation or be convenient for users. The Internet may be used in the transfer of information such as simulation models, data, results, and decisions among different simulation platforms or design stations. Simulation tools are bound to become more complicated and will require more technical know-how on the part of users to master the problem definition, simulation environment, and reporting details. The challenge of the next decade is to better integrate simulation in the building design process, for simulation tools to be more user-friendly and of a higher quality, and to exploit the explosion of communication opportunities that the Internet offers.

5 Conclusion

With the continuing effort to design and construct high-rise intelligent and sustainable building complexes in Hong Kong, the demand for advanced simulation tools to help project decisions will increase. Compared with the laboratory testing approach, computer simulation offers the advantages of fast, thorough and economical output based on a representative virtual environment.

Different kinds of building performance evaluation tools have been introduced in this chapter, together with application examples for illustration. While the tools are becoming more and more complicated and extensive, future development is still governed by the basic need for better support of the professional design and construction practice and better quality control over building performance assessments. Until now, most advanced predictive tools are regarded by building professionals as too difficult to learn, and can only be mastered by specialists or researchers. Insufficient user-friendliness and over-demanding user-training requirements have been the major barriers in promoting the wider use of these predictive tools.

In the longer term, advances in computer technology will make building simulation tools profoundly comprehensive and integrative in performance evaluation. At the simulation core, the inter-related sub-programs will support multi problem definitions, parallel data processing and online information transfer. The provision of an extensive central database and the preservation of the integrity of the building and the service systems will allow rigorous mathematical treatments and hence improve the quality of the simulation results. The simulation programs will be easier to master and finally reach a stage where most users in the industry will feel comfortable working with them.

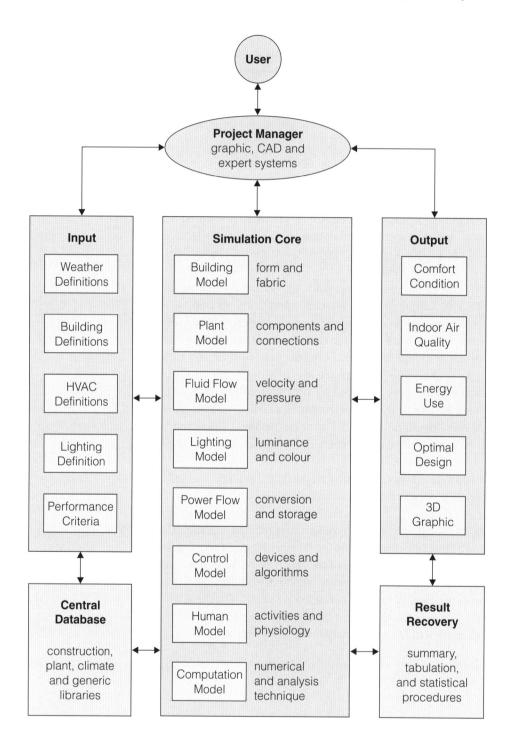

Figure 7.11   Features in ideal comprehensive building performance analysis software

References

1. American Society of Heating, Refrigeration and Air-conditioning Engineers (ASHRAE). 2001a. *2001 Fundamentals*, Atlanta, U.S. ASHRAE.

2. ASHRAE. 2001b. "Energy Standard for Building Except New Low-rise Residential Buildings." (90.1–2001). Atlanta, U.S. ASHRAE.

3. Chow, T. T., Chan, L. S., Fong, K. F., Yau R., Au, W. H., and Cheng, V. 2001. "District Cooling System Energy Modeling Methodology for New Urban Developments." Proceedings of the 4th International Conference on IAQ, Ventilation and Energy Conservation in Buildings (IAQVEC) 2001. Chengsha, China. 3: 1847–1855.

4. Chow, T. T., Lin, Z. and Liu, J. P. 2002a. "Effect of Condensing Unit Layout at Building Re-entrant on Split-type Air-conditioner Performance." *Energy & Building*. 34(3): 237–244.

5. _____. 2002b. "Effect of Condensing Unit Operation on Kitchen Exhaust at Residential Tower." *Architectural Science Review*. 45.1: 3–11.

6. Clarke, J. A. 2001. "Integrated Building Performance Simulation," Proceedings of IAQVEC 2001. 3: 1395–1404.

Chapter 8

The Tendering Systems of Public and Private Clients in Hong Kong

Ellen LEE

Tendering systems adopted by public and private clients are subject to different criteria, depending on the structure of the client firm and the nature of the projects. These criteria affect the content of the tender documents used between the client and the main contractor, as well as those between the main contractor and the sub-contractors. This chapter examines tendering procedures in Hong Kong, the criteria leading to the choice of tender documents, and their impact on pricing. The reader is required to have a basic understanding of different tendering methods, contractual arrangements, and procurement methods.

PART II
CHAPTER

8

The Tendering Systems of Public and Private Clients in Hong Kong

1 Introduction

In Hong Kong, a client who wishes to build a building would firstly appoint an architect to prepare a preliminary design or schematic design. If the client decides to proceed with the project, the architect will be asked to produce further drawings and/or specifications and quantity surveyors will be asked to prepare bills of quantities.

On the basis of the drawings, specifications and bills of quantities or similar cost documents, contractors will be invited to submit competitive tenders or enter into negotiations for carrying out the work. Usually the contractor who has submitted the lowest-priced tender is selected. This system has been in use since the turn of the century and is still adopted today. Although more design-and-build contracts have been let since the 1990s and partnering has been gradually adopted from 2000 onwards, the tendering systems in use have remained the same and are still based on competition and negotiation. Be it public or private clients, contractor selection criteria have changed little following these developments.

2 Tendering in the Construction Industry

The conventional separation of the design and construction processes reflects the methods of procurement, which is a process involving the selection of a contractual framework for a construction project. This invariably affects the choice of tendering methods and the types of tender documents used.

Hillebrandt (1991) identified four factors affecting the way in which tendering is carried out, how criteria are set and how documentation is used in the construction industry which distinguish it from other industries. These four factors are:

a)   The physical nature of the product;
b)   The structure of the industry together with the organization of the construction process;
c)   The determinants of demand; and
d)   The method of price determination

Each construction project is unique, which makes price comparison difficult. This is why standard documents or yardstick has to be used to provide a uniform basis for pricing. Thus, the tendering procedure and the content of the tender documents are equally important for selecting the contractor.

3   The Tendering Process

The tendering process begins with an invitation to tender and finishes with the appointment of a contractor to carry out the work. The client or the consultants on his behalf passes the tender documents to the contractor to price. The contractor fixes a price for the tender at which he is prepared to carry out the project or proposes a pricing method for consideration. If the client accepts the offer without any qualification, it becomes a legally binding agreement and the tender document is converted into a contract between the client and the contractor.

There are mainly two ways of selecting a contractor, by competition and by negotiation. While single-stage selective tendering is used for most projects, two-stage tendering is used for complex projects where contractors are shortlisted through some form of prescribed criteria before being selected to offer a price for competition. The offered price in such tender competition usually follows the traditional assumption that the lowest price is the key to acceptance of a bid, unless the offered price is based on negotiation with a single contractor.

Contractors usually use either one of the following two methods for determining the cost of construction works, namely the measurement method and the cost reimbursement method. These methods form the basis of the pricing documents, while the price, together with the technical capability of the contractor, forms the basis of the client's selection criteria. Most clients, particularly public and corporate clients, favour tendering by competition, while negotiation is favoured by contractors. Two main reasons are accountability to stakeholders and price variations.

Tendering is needed because there is no standard price for a piece of construction work. Each construction project is unique and the price indicator or cost yardstick is only indicative and not completely accurate. Cost experts can verify that a price is reasonable but not whether the price is exactly right. Since it is about handling future costs, a variance of +/– 5% is already regarded as highly accurate; a variance of +/– 15% is regarded as normal.

The commonest tendering methods are open tendering, selective tendering and negotiated tendering. Open tendering is no longer favoured by government departments and most public corporations in Hong Kong, and is now in limited use, mostly by some charity organizations or institutions which have restrictive funding. Selective tendering is the mostly used tendering method and can come in the form of a single-stage or two-stage system. In a single-stage tendering system, contractors are internally shortlisted before tendering and the contractors only compete on price. In two-stage tendering, the technical expertise and financial capabilities of the contractors are assessed in stage 1 or the prequalification stage of competition while the price competition is carried out in stage 2.

4 Criteria Applicable to the Form of Tendering

The choice of a correct method for contractor selection and the types of tendering documents used for price determination can be influenced by the following factors. A rational choice is one that best serves the client's interest even though it may be against the client's own judgement or preferences.

4.1 Size

Small projects are not suitable for the more elaborate forms of contractual arrangement. The resulting tendering procedures are also unlikely to be cost-effective. As small projects are more likely to adopt selective or open tendering or a design-and-build method, it is to the client's interest that the contractor prepares the pricing documents. Medium to large projects are suitable for a whole range of methods, although the traditional bills of quantities to be adopted as the tendering document are still favoured by most. With very large projects, when more advanced and complex forms of procurement may be necessary, the tendering system or the tendering document used will be dependent upon the procurement method adopted.

4.2   Client

Clients who regularly carry out construction work are better informed and staffed with knowledgeable personnel. They tend to have developed suitable and appropriate methods in-house and adopt procedures which are better suited to their company structure. Thus they tend not to require the same level of advice as those who only build occasionally.

The biggest client in Hong Kong is the government or the public client. It generally accounts for 40–50% of the total expenditure on building and construction. The operation of the World Trade Organization (WTO) system pre-sets rules for public procurement, requiring selective competition even when the choice is against its best interest. Private clients or commercial clients, on the other hand, have more autonomy in choosing the form of tendering that is most relevant to their specific needs.

The next biggest clients are the private sector property developers. Most of these large developers have contracting specialists, sub-contracting and suppliers' subsidiaries that form a vertical chain which supports the activities of their parent companies. These corporate clients may use negotiated tendering for projects to be carried out by companies under the same umbrella. Tender or pricing documents are prepared by in-house staff or outsourced to external consultants. For private sector clients who build for their own use, the user's participation in the project management process is very important in ensuring the client's needs are met.

4.3   Cost

It is generally believed that open tendering obtains the lowest price from a contractor. Negotiated tendering supposedly adds about 5% to the contract sum. Projects with unusually short contract periods tend to incur some form of cost penalty. The introduction of conditions that favour the client, or the imposition of higher standards of workmanship, also pushes up costs. Clients are more concerned with achieving an accurate prediction of cost at the earliest possible date and the lowest final cost. Of less importance are cash flow projections and indeterminate life-cycle cost. This would only be considered when a client is working with a longer-term perspective and has the intention to maintain a reputation alongside a corporate objective in the property development business. Whichever method is adopted, contractor selection is evaluated in the context of the characteristics unique to each project.

4.4 Time

The expected time of completion depends upon various factors. Factors to be considered include the shortest possible time from inception to handover, the shortest period of time when the contractor is actually on site (which affects overhead costs), reliable guaranteed completion dates, and phased completion to fit in with the client's project management programme. For instance, the time factor is critical for a user client who builds a new school and requires the building to be ready at the beginning of term, or a client building speculative housing units who requires the building to be ready for sale at an attractive price.

5 The Tender Document Used for Pricing

The tender document includes a section with either instructions for tenderers or information on the rules and criteria governing the tendering exercise that will influence how the works are priced. The price or agreement obtained this way will affect the choice of construction procedures and post-contract administration, particularly when disputes arise.

Based on the common procurement methods such as traditional bills of quantities and design-and-build used by public clients, the parts of the tendering document that determine the price for competition adopted by the Works Bureau of the Hong Kong government are:

a) Lump sum based on bills of firm quantities;
b) Lump sum based on drawings and specifications (sometimes with schedules of rates and approximate quantities submitted by tenderers);
c) Remeasurement based on bills of approximate quantities;
d) Remeasurement based on schedule of rates (applicable also to term contract);
e) Design and build based on target cost or cost reimbursement plus percentage/fixed fee.

Design-and-build contracts are traditionally lump sum contracts based on drawings and specifications prepared by consultants who are employed by the clients and who have the design responsibility. Nowadays, design-and-build has become a non-traditional procurement method where the design responsibility is shifted to the contractor. This change led to the publication of a new standard form of contract by the Hong Kong government in 2000 for use in conjunction with tendering for design and build.

5.1   Lump sum based on Bills of Firm Quantities

Both the quantities and unit rates in the bills of quantities are fixed. Bills of quantities provide a detailed breakdown of the tender sum and also a sound basis for the valuation of any variations to the design at the construction stage. Since the quantities are firm, it bears a fixed price (lump sum) on the tender. The client is thus made aware of his financial commitment for the purpose of obtaining adequate funding or prior approval for the proposed expenditure before the actual commencement of work. The bills of quantities are best used when there is adequate time for the preparation of a complete design prior to tendering arrangements, so that the quantities can be accurately measured before then.

5.2   Lump sum based on drawings and specifications

Tenderers prepare their own quantities from the drawings and specifications provided without bills of quantities supplied. This procedure is intended to be used for relatively small works and for specialist sub-contracting works. Design-and-build contractual arrangements incorporate a modified form of this, leaving the contractor to decide how to measure the works so as to arrive at a tender price for the works. The tender documents are sometimes supplemented with a schedule of rates for measuring variations or a schedule of quantities for comparing the basis of pricing when competition is keen.

Tenderers are supplied with complete working drawings and full specifications, while the design must be completed before the beginning of the tendering process. The drawings and specifications are used as tender documents when there is a lack of time and the client considers the benefits of bills of quantities to be less important than the early completion of the construction work. The advantage of having a lump sum is retained. When the selected contractors are specialists with the appropriate skills and reputation for the work to be carried out, a client may have to trust them not to form a tendering ring and overprice the work. A monitoring mechanism may be needed.

5.3   Remeasurement based on Bills of Approximate Quantities

The rates provided in bills of approximate quantities are fixed but the quantities are approximate and subject to re-measurement when the work is done. There are two possible reasons for the use of approximate quantities as a tender document:

a) The extent of the work is not known and the quantities cannot be accurately measured, although the details of the work can be defined. This normally applies to work related to ground conditions where a lot of site decisions have to be made as to the quantities of work to be carried out; or to repair and maintenance work where the extent of work to be carried out is uncertain at the tendering stage.

b) When the design is fairly well advanced but there is insufficient time to obtain accurate quantities, or the design is not sufficiently complete for that to be done. The bills of approximate quantities is also used when the client wants to have the advantage of detailed bills when work starts, without committing sufficient time for the preparation of a bill of firm quantities. This is not uncommon where early completion is required and time is a critical factor, but the client runs the risk of low predictability of the final cost. It is an alternative which reduces planning time and allows a client to go for tendering sooner than normal, but the client will be less certain of his eventual financial commitment at the tendering stage and resources have to be provided for the remeasurement process.

5.4   Remeasurement based on Schedule of Rates

There are two forms of schedule of rates used in tender documents: standard rates and ad-hoc rates:

Standard rates

A standard schedule of rates lists under appropriate trade headings all the items likely to arise in any construction project, with a unit rate against each item. The government standard is the Schedules of Rates for Maintenance and Minor Works which are used when term contract is the preferred contractual arrangement. A term contract is a contract specifying the period of time during which the contractor is responsible for the works listed or ordered for execution. A schedule of rates is an appropriate pricing document for a term contract because there is a limited range of work of a repetitive nature to be carried out. Tenderers are asked to submit percentage additions or deductions to the listed rates, usually by sections or sub-sections covering different trades, thus allowing for variations in construction costs which reflect the time value of cost since the date of the preparation of the schedule.

"Ad-hoc" rates

This is a schedule specially prepared for a particular project and lists only those items which are appropriate to that project, including any special or unusual items. An "ad hoc" schedule may be pre-priced or left blank for the tenderers to insert individual rates against each items. The latter method makes the comparison and assessment of tenders much more difficult because of the absence of quantities. The major merit of this method is that the items are "made to measure" and tenderers are only required to concern themselves with a restricted range of items, thus enabling them to assess rates or percentages more accurately.

Cost reimbursement or prime cost

Prime cost refers to the actual cost of buying materials, goods and components, using or hiring plant and employing labourers to carry out construction work, plus a fee covering the contractor's overheads and profit. The various forms in use are cost plus percentage, cost plus fixed fee and target cost.

This method of pricing produces the most uncertain financial outcome. Tenders using this kind of pricing method contain no total sum and it may be very difficult to give any reliable estimate of the final cost. Their appeal to clients is that the work on site can commence in the early stages of design, when the earliest possible commencement time may be financially more advantageous than a lower final cost of construction.

Target cost

Target cost is now gaining popularity in the U.K. because of the practice of partnering. The target cost is an estimate of likely total cost; i.e. the contractors tender on the basis of a guaranteed maximum price which allows the client to know the possible maximum price at completion. In view of the trend towards a partnering approach, target cost may be a suitable option for tender documentation.

6   The Tendering System of Public Clients

The Hong Kong government has adopted the procurement methods of the WTO for the purchase of materials and supplies, and this also applies to procurement for construction work and consultancy contracts. Potential tenderers bid according to the terms and conditions established by the government for each category of works.

6.1 Works Bureau rules of tendering

The government as a public client has detailed and comprehensive tendering rules and procedures for contractor selection. Public works are divided into seven categories (Table 8.1). Each construction work category is managed by a department within the government, but that managing department is not necessarily responsible for servicing and monitoring the performance of the contractors for that category of works. To be qualified to tender for public works, contractors have to apply to the Works Branch for admission to the List of Approved Contractors for Public Works (the List). The contractors will then be judged on their financial, technical and management capability for eligibility to tender for certain category of works. The works categories are divided according to Group Tender Limits based on contract value and certain financial criteria (Table 8.2). For application for admission to the List, contractors have to produce their latest unconsolidated audited accounts together with their evidence of technical and management capabilities.

Construction Work Categories	Managing Department
Buildings	Architectural Service Department
Port works	Civil Engineering Department
Roads	Highways Department
Site formation	Territory Development Department
Drainage works	Drainage Services Department
Electrical & mechanical works	Electrical & Mechanical Services Department
Waterworks	Water Supplies Department

Source: Works Bureau, Hong Kong Special Administrative Region Government.

Table 8.1   Public works categories

For the Buildings category, once the application for admission has been approved, a contractor has to register with the Building Authority in accordance with the Buildings Ordinance (Cap 123). His name is then published in the *Hong Kong Government Gazette* along with the respective category/categories, groups and status for which he has been qualified. However, the approved contractors may have to go through a minimum probationary period of two years before their status is confirmed. Contractors do not have to start at Group A and proceed upwards, but a contractor confirmed for Group A can ask for promotion to the next higher group in a particular category provided he meets the financial criteria and has the technical and management capabilities required for the higher group.

Group Tender Limits	Contracts of Value (as at 9/97)	Financial Capability	
		Minimum Employed Capital (HK$)	Maximum Working Capital (HK$)
Group A	Up to $20 million	1,100,000 (P) 2,100,000 ©	1,100,000 (P) 2,100,000 ©
Group B	Up to $50 million	2,600,000 (P) 5,300,000 ©	2,600,000 (P) 5,300,000 ©
Group C	Exceeding $50 million	7,100,000 (P) 9,000,000 ©	7,100,000 (P) 9,000,000 ©

(P) Probationary; © Confirmed

Source: Works Bureau, Hong Kong Special Administrative Region Government.

Table 8.2   Contractor groups based on contract values and financial capability

Public clients shortlist contractors as part of a prequalification process. The classification of Groups A/B/C is based on the value of the project the tenderers are qualified to bid for. Prequalification is necessary when the scope of work of a new project is complex and unique. According to the Works Bureau, this normally applies to projects of an unusually high value which are subject to time constraints and requires a high level of co-ordination and special technical expertise. Non-traditional contractual arrangements may be used in such cases, but priority will be given to contractors in the approved list. However, under special circumstances, opportunity will be given to other suitable local and overseas contractors not on the List. When a project is open also to overseas contractors, the government will place notices in overseas publications and government representative offices in addition to the usual local notification.

Generally six months are allowed for the prequalification exercise and contractors' views will be considered with regard to enhancement of the tender documents. There is no limit to the number of contractors applying for prequalification, which comprises two stages of assessment: (1) inadequate contractors are screened out; and (2) the remaining qualified contractors are evaluated on their management and technical capabilities.

The stage 1 assessment is based on the type and value of the contracts the contractors have undertaken during the last two years, their financial capability and the availability of special plants and equipment for use on the project. The stage 2 assessment employs a marking scheme with relative weightings which gives an eligible contractor a maximum of 100 points and a minimum level for acceptance under each selection criteria (Table 8.3).

Aspects		Maximum Mark
(1) Contractor's experience	Relevant experience in contracts of similar type or size in the past 5 years Relevant local experience	15
(2) Contractor's past performance (technical)	Workmanship Progress Claims attitude	15
(3) Contractor's past performance	Site safety Environmental performance Care of utilities and the general public Compliance with regulations	10
(4) Contractor's resources	Managerial, technical staff Plant and equipment	20
(5) Contractor's technical ability	Demonstration on site, method statement, quality assurance plan, site safety policy etc.	40
		100

Source: Works Bureau, Hong Kong Special Administrative Region Government.

Table 8.3 Marking scheme for assessment of contractor's management and technical competence

The contractors have to complete a prequalification document for stage 2 of the evaluation. Prospective contractors may be asked to submit further information or attend an interview if the information is insufficient for making a judgement. Stage 2 of the evaluation (Table 8.4) covers most of the following:

a) Experience in government contracts, local contracts and overseas contracts in the past five years;

b) Project experience to verify type, size and form of construction in the past five years;

c) Availability of special plant, equipment and workshops;

d) Proposals on preliminary method statement, quality assurance plan and site safety policy;

e) Site safety record;

f) Financial resources;

g) The contractor's brief history of claims, litigation and arbitration on government contracts in the past five years;

h) Past environmental performance;

i) Conviction records, if any, related to construction.

Contractors' performance will also be monitored through regular reports (Appendix I) on running projects. Reviews are carried out to maintain acceptable performance and financial status of contractors. Such reports will be kept in a Central Record System in the Works Bureau to be readily available for tender assessment.

Contract information	Information required from prospective tenderers	Specific queries

Description of the works
Project team
Tender documentation
Tender programme
Construction programme
Other information provided
    (e.g. contractor's design liability,
    works by specialist contractors
    and nominated sub-contractors)
Rejection of application and
    disqualification of tenders
Joint ventures
Selection criteria
	Details of applicants
	Experience
	Resources
	Sub-contracts
	Project management
	Site safety
	Preliminary method statement
	Joint ventures
		Any particular issues
		to be addressed

Source: Works Bureau, Hong Kong Special Administrative Region Government.

Table 8.4   Scope of prequalification document

Disqualifying contractors

Contractors can be removed from the List, suspended for a period of time, downgraded to probationary status or demoted to a lower group for the following reasons:

a) Unsatisfactory performance
b) Serious misconduct or suspected serious misconduct
c) Insolvency or other financial problems
d) Poor site safety record
e) Poor environmental procedure
f) Convictions
g) Failure to employ the minimum number of full-time management and technical personnel
h) Violation of laws

Contractors traditionally compete on price with the lowest-priced tenderer winning the contract. This has occasionally led to low quality work and non-performance, as a result of which contractors are now required to compete on performance as well as price.

6.2   Consultancy tendering

In addition to the tendering system for contractor selection, public clients also have a competitive fee tendering system for selecting professional con sultants such as architects and quantity surveyors. This is a way of better utilizing their internal resources in circumstances where the workload outweighs their internal capacity. A recent survey of Hong Kong Quantity Surveyor (QS) Consultants shows that 89% of QS consultancy work is allocated through competitive tendering (Drew 2000).

7 Architectural Services Department Practices

The Architectural Services Department (ASD) is the department which handles most of the building construction works for the government. Following the Works Bureau rules, a pre-qualified tenderer or a listed tenderer places the technical submission and the tender price document into two separate envelopes marked "Technical Submission" and "Tender Price Documents" respectively. The technical submission is for the assessment of the contractors' capability and would not form part of the contract. A conventional single-stage selective tendering system is adopted for the construction of standard projects but has its own guidelines for contract awards covering the status of the tenderers, the number of tenderers in competition and the shortlisting criteria for the selection of contractors.

It is common in Hong Kong for a contracting company to be the subsidiary of a holding company. The tenderer can be a holding company or a joint-venture company, but a holding company and its subsidiary cannot tender for the same job at the same time. The rules recommend that the number of tenderers should be limited, although up to 30 may be invited to tender. The number of tenderers is normally restricted to 10–12. The shortlisting criteria include the firm's financial standing and record; recent job experience involving similar contract periods; general experience and reputation of the firm for similar building types; adequacy of management; and adequacy of capacity.

A standard form of preliminary enquiry containing information of contractual significance is sent to tenderers to solicit their willingness to tender. Once a contractor has confirmed an intention to tender he would normally receive the tender document on the same day. The rules assume the use of a standard form of building contract with which the parties involved are familiar.

7.1 The submission

If a tenderer submits a qualified tender, an opportunity would be given to withdraw the qualification without amending the tender figure, otherwise the tender would normally be rejected. A tender may be withdrawn at any time before acceptance. After the tenders are opened, all but the lowest three tenderers will be informed immediately, and the tendering result will be released once a contractor is selected.

7.2   Pricing errors

The rules set out ways of dealing with errors. Normally the tenderers are given an opportunity to confirm the offer, whether there are errors to be corrected or not, or withdraw the offer. Where the tender sum is corrected and is no longer the lowest, then the next lowest tender would be considered. Where the offer is confirmed an endorsement would be added to the priced bills such that all rates, except preliminary items, contingencies, prime cost and provisional sums are reduced or increased, as appropriate, according to the same adjustment percentage of the corrected total.

In Hong Kong, tendering procedures normally contain clauses governing how errors should be dealt with. Usually, the tendered sum is not allowed to be amended for errors found during the examination of tenders. Knowledgeable clients such as the government tend to have the tender sum fixed at the time the tender is opened, to simplify administrative procedures, and contractors have to bear the risk if they have made any mistakes in the tender. The risk of errors in pricing therefore rests completely with the contractors, who have to exercise care in preparing the tender estimate. Should examination of a tender reveal errors of such magnitude as, in the opinion of the architect, to cause the contractor a serious loss (e.g. exceeding 2.5 % of the tender sum), then the nature and amount of such errors will be communicated to the tenderer and he will be asked to confirm in writing that he is prepared to abide by his tender or that he wishes to withdraw his tender. The local practice is very similar to the code of procedure in the U.K.

7.3   Reduction bill

If the tender under consideration exceeds the estimated cost, negotiations would take place with the tenderer to reduce the price. The quantity surveyor would then produce what is called "reduction or addendum bills" that are priced and signed by both parties as part of the contract bills. For accountability reasons, some corporate clients (e.g. government depart-ments or similar organizations) are kept away from the negotiations and have to either accept or reject the tendered price. When these negotiations fail further negotiations may proceed with the next lowest tenderer.

8   Housing Department practices

The Housing Department working under the Housing Authority started a quality assurance system called the Performance Assessment Scoring

System (PASS) in 1990 to assess the performance of contractors under a scoring system. The contractors to be assessed must first of all be one of the listed contractors. After inspection and checking, each contractor is given a score indicating the level of their performance on a project. This will entitle the contractor to tender for further housing projects. However, if the score is below the required level, they will not be allowed to tender until they improve their scores. This is to maintain not only the standard of entry, but also the quality of work.

PASS sets down the admission criterion for contractors seeking to tender for housing jobs. The PASS manual (1994) provides information on the three areas assessed, i.e. input, output and the maintenance period.

8.1   Assessment areas

Input assessment is conducted quarterly through formal meetings attended by the project team including the client, the consultants and the main contractor based on the contractor's current track record in the following areas:

a) Management and organization of works
b) Resources
c) Coordination
d) Documentation
e) Programme and progress
f) Completed works after sectional completion

Output assessment consists of an inspection and record check in the following areas:

a) Structural works
b) Architectural works
c) Other obligations

It is conducted monthly by project teams with the assistance of site staff, in the presence of the contractor's representative.

Maintenance period assessment is conducted quarterly to assess the contractor's performance during the maintenance period in the following areas:

a) Outstanding works
b) Execution of repair works
c) Management, response and documentation

Input and output assessment will together form a composite score at a ratio of 1:3. The assessment is continuous and the contractor's performance is monitored regularly throughout the project. The frequency of assessment varies depending on the status of the project. For maintenance assessment, domestic buildings are assessed at monthly intervals while non-domestic buildings are assessed at quarterly intervals.

8.2   Alternative tender

When the original tender (called Tender A) is submitted with qualification, it will be regarded as an unauthorized qualification for the tender and this will cause the tender to be disqualified. However, tenderers are allowed to submit an alternative tender (called Tender B) usually for a shorter construction period. Tenderers also have to submit a method statement including details concerning the viability of the change. This working method makes comparison feasible in forming an equal basis for tendering and makes selective tendering meaningful.

8.3   Tender evaluation criteria

Tender Limit	The Recommended Tender
For tenders less than $10 million	within $1 million of the lowest tender
For tenders over $10 million	within 10% of the lowest tender
For tenders outside the limits mentioned above	endorsement by the appropriate Director or delegate is necessary

Source: Hong Kong Housing Authority.

Table 8.5   Recommended value for housing projects

9   The Tendering System of Private Clients

Private clients who want to prequalify contractors adopt a procedure similar to that of a tendering procedure where contractors are asked to submit information about their company background, project records, financial materials, project personnel (staff's experience and qualification), etc. This can be referred to as two-stage selective tendering where the contractors are selected first before they are asked or invited to price the tender.

Private clients generally adopt the code of tendering procedures published by NJCC (National Joint Consultative Committee of Architects, Quantity Surveyors, and Builders) of the U.K. The procedure may be modified to suit the structure of the company and local culture and practice. The purpose of tendering is to maintain openness and fairness.

However, private clients have more flexibility than public clients as the former only have to take their own criteria into account. When the client is a listed company or a corporation, then they have to be accountable to their stakeholders and flexibility will consequently be restricted.

Private clients' requirements in terms of their company structure or the project itself are given in the form of instructions to tenderers. In principle, there are no rigid rules governing contractor selection by private clients and the code of procedures adopted by professional quantity surveyor is generally applicable. The usual practice is to draw up a shortlist of prospective tenderers based on their knowledge, experience in working with the client or a third party reference. The selected list varies from firm to firm and from project to project; the criteria for selection are generally based on the recommendation of the professional surveyor or consultant. They are similar to the procedures adopted by public clients, but with more flexibility in setting up the selection criteria.

10 Conclusion

Tendering by competition appears to be the norm. Public clients are restricted to tendering by competition in following the WTO's procurement system. Tendering rules so set therefore dominate how government projects are let. Private clients' autonomy is dependent on how "private" the company is and the number of stakeholders they are responsible for. So a listed company or a corporate client would have more accountability and less flexibility than a limited company or an individual client. The question then is, given the choice, would a private client go for tendering by negotiation, and not by competition? The major concern for negotiation may not be the non-competitiveness of the price, but may rather be a kind of personal favour or some benefits in the transaction not accepted by the stakeholders. Negotiation, given its own merits, is viewed by most as possible only when organizational affiliation exists between the client and the contractor, such as parent-subsidiary companies, partnership, joint-venture or alliances, if the work does not arise from emergency reasons.

The prequalification process is part of a quality assurance system used to ensure a suitable and capable contractor is selected to construct a project. Although contractors are assessed in terms of their technical submission as well as price, there is a tendency for the price to dominate as the contractors prequalified to tender may not be very different in terms of technical capability. The weighting assigned to each factor also has a role to play. If the technical submission to price weighting is 20:80 or 30:70, price will dominate the selection and tenderers will essentially compete on price (the lowest tender will be accepted). The Hong Kong Housing Authority Consultative Document on Quality Housing (Jan. 2000) and the Report of the Construction Industry Review Committee (Jan. 2001) recommend that the weighting should be altered to 50:50 or an even higher weighting be given to the technical submission. Until a comprehensive system feasible for public works or semi-public works tendering is devised, such a change may be difficult to implement due to public accountability.

The present tendering system has worked well in terms of comprehensiveness and the fairness it promotes. But China's open-door policy has given rise to a dilemma concerning the competitiveness of pricing as similar services can be purchased at a comparatively lower price across the border. This competition is particularly severe with the increased ease of global communication and the anticipated new form of documentation transfer. Only when price becomes less emphasised in the selection process would contractors or whoever in the tendering channel be adequately motivated to do a good job.

References

1. Drew, D. 2000. "Analyzing Competitiveness with a View to Enhancing Consultants' Fee Tendering." Paper presensed at a seminar organized by the Division of Building Science & Technology, City University of Hong Kong.

2. Environment, Transport and Works Bureau. www.etwb.info.gov.hk.

3. _____. 1996–2001. Technical Circulars. Hong Kong Special Administrative Region.

4. Hong Kong Housing Authority. 1996–2000. Contract and Tender Procedures Manual. Hong Kong Special Administrative Region.

5. _____. 2000. "Quality Housing: Partnering for Change." Consultative Document. Hong Kong Housing Authority.

6. Kwakye A. A. 1997. *Construction Project Administration in Practice.* Longman.

7. National Joint Consultative Committee. 1977. *Code of Procedure for Single Selective Tendering.* London: RIBA Publication.

8. _____. 1983. *Code of Procedure for Two Selective Tendering.* London: RIBA Publication.

9. Ramus, J. & Birchall, S. 1996. *Contract Practice for Surveyors.* 3rd edition. Oxford: Butterworth-Heinemann.

10. Rowlinson, S. M. & Walker, A. 1995. *The Construction Industry of Hong Kong*. Hong Kong: Longman.

11. Tang, H. 2001. "Construct for Excellence." Report of the Construction Industry Review Committee.

12. Willis, C. J., Ashworth, A. & Willis, J. A. 1994. *Practice and Procedure for the Quantity Surveyor*. 10th edition. Blackwell Scientific Publication.

Appendix I
Reporting Form Used for Monitoring Contractor's Performance

REPORT ON CONTRACTOR'S PERFORMANCE

PART I - SUMMARY OF PERFORMANCE

DEPARTMENT/OFFICE :-

REPORT FOR QUARTER :-   (   ) Quarter/Period/Maint./Final *

REPORTING PERIOD :

From :-   To :-

A   Contractor's Details

Contractor's Ref. :-

Contractor's Name :-

Contract No. :-   PWP No. :-

Contract Title :-

Type of Contract :-   Civil / Building / Term / Specialist / Maintenance *

Predominant Category of Work :-

Tenders invited from

List/Group :-   Cat :-   Prequalified :-   Y/N *

Contractor's Classification

List :-   Group :-   Cat :-

B   Contract Stage and Duration

Commencement Date :-

Original Contract Completion Date :-

Original Contract Period :-   months

Completion Dates+			
Section	Original	Extended	Anticipated/Certified
Whole			

Contract Stage *

1. Design & Submission of Drawings
2. Under Manufacture
3. Under Construction/Site Work Under Progress
4. Substantial Completion
5. 6 Months after Substantial Completion
6. Defects Liability Period
7. Maintenance Certificate Issued
8. Final Completion Certificate Issued

C   Contract Value

Original contract sum	$M	
Estimated value of work done for this quarter	$M	
Estimated value of work done to date	$M	
Estimated value of work outstanding	$M	
Other matters allowed for in final contract sum	$M	
Estimated final contract sum	$M	

D   Performance

		VG	S	P
1	Workmanship			
2	Progress			
3	Site safety			
4	Environmental pollution control			
5	Organisation			
6	General obligations			
7	Industry awareness			
8	Resources			
9	Design			
10	Attendance to emergency			
	Overall performance			

VG :- Very Good
S :- Satisfactory
P :- Poor

Note:

A "Poor" in any one of sections 1, 2, 3, 4 will be a mandatory "Poor" in the "Overall performance" and the report will be rated as "Adverse".

Appendix I

Reporting Form Used for Monitoring Contractor's Performance (cont'd)

E Claims #	No.	Claimed	Assessed	Awarded	Awarded/Assessed	Unresolved
EOT claims						
Up to last period		days	days	days	%	days
Total to-date		days	days	days	%	days
Monetary claims						
Up to last period		$M	$M	$M	%	$M
Total to-date		$M	$M	$M	%	$M

Attitude to claims : reasonable/unreasonable *   (If unreasonable, comment)

F   Remarks by Reporting Officer

(Include an assessment of the contractor's overall progress compared to his original programme, see Section 2 of GN. The Chief Engineer/Architect/Head of Office (for Consultants Administered Projects) should entirely satisfy himself/herself that there is adequate documented evidence to prove that an Adverse report is warranted before endorsement.)

Report   NOT ADVERSE / ADVERSE *

Reported by :   Agreed by :   Endorsed by :

(_____)   (_____)   (_____)
Engineer/Engineer's/   Senior Engineer/Engineer/   Chief Engineer/Architect/
Architect's * Representative   Architect * for the Contract   Head of Office (for Consultants
for the Contract   Administered Projects) *

Date ....   Date ....   Date ....

G   Remarks by Reporting Review Committee :-

(The Chairman of the Reporting Review Committee should entirely satisfy himself/herself that there is adequate documented evidence to prove that an Adverse report is warranted before endorsement/amendment.)

This report is endorsed/has been amended * by the Reporting Review Committee
Remarks (only if Engineer/Architect report amended) :-

                                        ....
                                        Chairman, Reporting Review Committee

Report   NOT ADVERSE / ADVERSE *   Date ....

Number of site instructions issued to-date :

(*)   Delete as appropriate
(+)   All dates to be shown as dd/mm/yy e.g. 31/10/93 (add suffix (A) or (C) after "Anticipated/Certified")
(#) 1. Claims for monies for measured or varied work, star rates, omitted items etc. should not be treated as a claim.
        2. The number of claims entered should not be the number of notices of claims received but should be the number of claims quantified.

Appendix I

Reporting Form Used for Monitoring Contractor's Performance (cont'd)

REPORT ON CONTRACTOR'S PERFORMANCE				
PART II - INDIVIDUAL ASPECTS OF PERFORMANCE				
ITEM	ASPECTS OF PERFORMANCE	Very Good	Satisfactory	Poor
Section 1	Workmanship			
1.1	Standard of temporary works			
1.2	Standard of materials/equipment supplied			
1.3	Standard of workmanship, earthworks			
1.4	Standard of workmanship, structural			
1.5	Standard of workmanship, finishes			
1.6	Standard of workmanship, conduit/pipe/duct works			
1.7	Standard of workmanship, equipment and plant			
1.8	Standard of workmanship, testing and commissioning			
1.9	Standard of workmanship, others			
	Overall rating			
Section 2	Progress			
2.1	Adequacy of programme			
2.2	Adherence to programme			
2.3	Updating of programme			
2.4	Suitability of method and sequence of working			
2.5	Achievement in period			
2.6	Action taken to mitigate delay/catch up with programme			
	Overall rating			
Section 3	Site safety			
3.1	Provision and maintenance of plant			
3.2	Provision and maintenance of working environment			
3.3	Provision of information, instruction and training			
3.4	Provision and implementation of safe systems of work			
3.5	Employment of safety officer/supervisor			
3.6	Site accident record			
	Overall rating			
Section 4	Environmental pollution control			
4.1	Adequacy of water pollution avoidance measures			
4.2	Adequacy of noise pollution avoidance measures			
4.3	Adequacy of air pollution avoidance measures			
4.4	Adequacy of waste pollution avoidance measures			
4.5	Compliance with environmental enactments			
4.6	Action taken to remedy non-compliance			
4.7	Implementation of waste management plan			
	Overall rating			
Section 5	Organisation			
5.1	Adequacy of organisation structure			
5.2	Support by head office			
5.3	Adequacy of planning of work			
5.4	Adequacy of executive authority			
5.5	Control of supervisory staff by Site Agent			
5.6	Management of sub-contractors by Site Agent			
5.7	Adequacy of site supervisory staff			
5.8	Identification of and responsiveness to problems			
	Overall rating			

Appendix I

Reporting Form Used for Monitoring Contractor's Performance (cont'd)

ITEM	ASPECTS OF PERFORMANCE	Very Good	Satisfactory	Poor
Section 6	General obligations			
6.1	Cleanliness of site			
6.2	Care of works			
6.3	Compliance with insurance requirements	■■■■		
6.4	Coordination of utilities and other authorised contractors			
6.5	Compliance with conditions on road openings			
6.6	Care of utilities			
6.7	Compliance with enactments other than environmental			
6.8	Adequacy/submission of operational and maintenance manuals			
6.9	Training of employer's personnel			
6.10	Adequacy of notice for inspection of works			
6.11	Payment of nominated sub-contractors	■■■■		
6.12	Compliance with particulars related to sub-contracting			
6.13	Attention to site security			
6.14	Attention to records			
6.15	Attention to submission of accounts/valuations			
6.16	Control of materials supplied by Government			
	Overall rating			
Section 7	Industry awareness			
7.1	Employment of technician apprentices, building & civil engineering graduates and ex-CITA trainee craftsmen			
7.2	Training of technician apprentices, building & civil engineering graduates and ex-CITA trainee craftsmen			
7.3	Care and welfare of workers			
7.4	Care of the general public			
7.5	Employment of qualified tradesmen and intermediate tradesmen			
	Overall rating			
Section 8	Resources			
8.1	Adequacy of plant resources			
8.2	Adequacy of labour resources			
8.3	Adequacy of material resources			
	Overall rating			
Section 9	Design (design & build contracts only)			
9.1	Collection & appreciation of information (e.g. design codes, design parameters)			
9.2	Design solutions (design analysis)			
9.3	Consultation with relevant government departments			
9.4	Quality and presentation of design			
9.5	Availability of design in time			
9.6	Quality of as-built drawings			
9.7	Availability of as-built drawings in time			
	Overall rating			
Section 10	Attendance to emergency (term contracts)			
10.1	Response to call			
10.2	Speed of emergency repairs			
10.3	Attendance to routine/on-call/emergency repairs			
	Overall rating			

Notes

1. Mark appropriate box of performance for each item with "x".
2. For items which are not applicable to the contract, put "N.A.".

Source: *Environment, Transport and Works Bureau, Hong Kong Special Administrative Region Government 2002.*

Part III
Construction Process

Chapter 9

Site Production Layout Planning for High-rise Building Construction

Arthur W. T. LEUNG

Site production layout planning is one of the most important construction planning activities in a building project since effective planning can improve productivity. The allocation of site facilities is a complicated task as building projects are unique. The nature and conditions of construction sites vary depending on site access, geographic profiles, building layout, and building shape and size. Construction planners, therefore, have to analyze site conditions and devise the optimum position of site facilities. This chapter discusses the scope, criteria, and strategies of site production layout planning with the help of an illustrated example.

9

Site Production Layout Planning for High-rise Building Construction

1 Introduction

Site production layout planning (SPLP), which is also known as site layout plan, is an important production planning activity in building construction. Efficient SPLP can improve productivity.

Production costs appear to be hidden costs in the mind of customers since they pay more attention to the functions and materials used for a product. In the manufacturing industry, when offering a competitive product to the market, a manufacturer will cut down the selling price in order to gain market share. Of all the cost items, production cost is the one over which a manufacturer can have relatively high autonomy in terms of cost control. The planning of factory layout plays an important role in minimizing the costs of production. The layout planning of a production line must take into account the co-ordination of work sequences in relation to the resources assigned, namely materials, manpower, and plant and equipment. The objective is to cut down costs by minimizing the production time. A factory production line can be defined as a permanent set-up whereas a site production line is a temporary set-up that will be dismantled upon completion of a project. In an automobile production line, the change of a car model may involve a change of the engine and the inclusion of the latest accessories. This will lead to the relocation of the processing points and storage points along the production lines. The core of the production process and work flow are maintained.

SPLP is one of the critical production planning processes in high-rise building construction. In site production, the processes involve both horizontal and vertical movements. In constructing a concrete floor, for example, starting from the ground floor, the production usually starts with the construction of the vertical members such as columns and walls, followed by the construction of the horizontal members, the beams and slabs. Subsequently, the production line has to be extended upwards for constructing the next floor until it reaches the roof level. Unlike factory production, site production lines have to be dismantled when the building is completed. Since a building design is normally unique, a site production layout is also unique for each building site. Construction sites located in the urban area impose additional constraints through lack of site space for setting up the site production line. This chapter discusses the SPLP principles for high-rise reinforced concrete building construction with emphasis on the construction of the concrete frame on confined sites.

2   Scope of Site Production Layout Planning

Cheng and O'Connor (1996) stated that temporary facilities (TF) are facilities located on site to support construction operations. An efficient TF site layout locates facilities such as job offices, warehouses, workshops, etc., to optimize workforce movements, material and equipment handling, travel distance, traffic interference, and the need for expansion and relocation. They also stated that the primary concern in site layout is to identify suitable areas within which to locate TFs.

Calvert, Bailey and Coles (1995) also stated that a site layout plan is a plan which should be drawn up showing the relative location of facilities, accommodation and plant, with the overall intention of providing the best condition for optimum economy, continuity and safety during building operations.

In order to assign priority in site layout planning, the scope of SPLP can be divided into production facilities and supporting facilities.

Production facilities can be defined as those contributing directly to the construction processes whereas supporting facilities provide indirect support to these processes. A tower crane that contributes to material transportation would be classified as a production facility. On the other hand, site accommodation such as offices, resting places, storage and working shelters for staff and workers would be classified as supporting facilities. However, there are grey areas in the classification of facilities. In classifying an access road, it is difficult to define its nature in terms of a physical production plant or a supporting feature. It will be more appropriate to further break down supporting facilities into vehicular facilities and logistic facilities. Table 9.1 shows the major temporary facilities for a typical site.

Production Facilities	Vehicular Facilities	Logistic Facilities
Tower crane	Site entrance	Offices
Material hoist	Periphery access road	Passenger hoist
Concrete batching plant	Internal access road	Plant workshop
Static concrete pump	Passing bay	Site laboratory
Steel bending yards	Loading and unloading bays	Common building materials storage yards
Formwork fabrication yard	Concrete loading and unloading points	Storage yards
Precast concrete yard	Carpark	Canteen
Scaffolding	Wheel washing bay	Toilet and changing room
		Guard room
		Refuse chute

Table 9.1   Classification of temporary facilities

Thus, an efficient temporary site layout would locate facilities such as offices, storage yards, fabrication and workshops, etc., to optimize workforce movement, material and equipment handling, travel distance, traffic interference, and the need of expansion and relocation.

According to the above definitions and classification, SPLP is basically defined as site space and process points allocation for materials storage, working areas, site accommodation, plant positions and general circulation areas within a site. These facilities are integrated to form a temporary production line for construction work and will usually remain on site for the duration of the project. As a result, optimal resources, time, cost and safety can be achieved through this planning exercise.

The aims of SPLP can be summarized as follows:

a) minimizing production time through reduction in material transportation time;
b) minimizing non-productive time through avoidance of double handling and congestion;
c) providing appropriate working conditions for site personnel;
d) providing appropriate storage yards for materials;
e) maintaining and promoting safety and health; and
f) providing adequate security to prevent loss of properties.

3   Priority in Site Layout Planning

Tommelein, et al. (1991) identified three stages in construction site layout:

a) identifying facilities that are temporarily needed to support construction operations on a project, but that do not form a part of the finished structure;
b) determining the size and shape of these facilities; and
c) positioning them within the boundaries of available on-site or remote areas.

For factory planning, production engineers will submit their production layout to an architect to produce a building design that is in accordance with their factory production requirements (Meyers 1993). It is presumed that sufficient land will be procured to accommodate the facilities. However, site space available for construction activities is limited and scarce. This imposes constraints on SPLP.

In solving SPLP problems, site planners have to prioritize the facilities. Leung, Tam and Tong (2001) conducted an opinion survey and generated a priority list for site facilities. An analysis of the ranking, nature and general siting positions of the facilities is shown in Table 9.2. Four out of the top five in ranking are production facilities. Tower cranes that provide three-dimensional transportation for materials are ranked first on the list. Material hoists that provide major vertical transportation for materials are ranked second. The concrete pump and concrete batching plant are also in the top five. The list shows the construction of the reinforced concrete frame for a high-rise building is the main focus of SPLP. Passenger hoists, which are logistic facilities, are ranked third on the list. The ranking is reasonable as the provision of the passenger hoist can minimize travelling time and unnecessary effort in reaching the workplace of workers or site personnel. This is particularly important when constructing a high-rise building of 20 floors or more. The logistic facilities and the vehicular facilities are spread quite evenly in the rest (below the top five) of the priority list, although the logistic facilities are generally ranked higher in the list.

The ranking also reveals that facilities usually attached or installed close to the building are ranked higher as they make a greater contribution to the production process. Facilities usually located along the site periphery are normally ranked at the bottom of the list. The priority in terms of location is:

a) attached to the building;
b) close to the building;
c) around the building; and
d) along the site periphery.

Ranking	Facility	Classification	Location
1.	Tower Crane	Production Facilities	Attached to the building
2.	Material Hoist	Production Facilities	Attached to the building
3.	Passenger Hoist	Logistic Facilities	Attached to the building
4.	Concrete Pump	Production Facilities	Close to the building
5.	Concrete-batching Plant	Production Facilities	Close to the building
6.	Refuse Chute	Logistic Facilities	Attached to the building
7.	Site Access	Vehicular Facilities	Around the building
8.	Site Office	Logistic Facilities	Along the site periphery
9.	Loading Point for concreting	Logistic Facilities	Close to the building
10.	Steel Bending Yard	Production Facilities	Close to the building
11.	Formwork Fabrication Yard	Production Facilities	Close to the building
12.	Plant Workshop	Logistic Facilities	Along the site periphery
13.	Toilet	Logistic Facilities	Along the site periphery
14.	Site Laboratory	Logistic Facilities	Along the site periphery
15.	Car Parks	Vehicular Facilities	Along the site periphery
16.	Canteen	Logistic Facilities	Along the site periphery
17.	Entrance	Vehicular Facilities	Along the site periphery
18.	Wheel Washing Bay	Vehicular Facilities	Along the site periphery
19.	Scaffolding	Production Facilities	Attached to the building
20.	Working Area for sub-contractor	Logistic Facilities	Along the site periphery

Table 9.2   Priority of temporary facilities

The logical sequence in allocating site space for temporary facilities can be summarized below:

a) allocating key production facilities that are to be attached to the building;

b) allocating the remaining production facilities close to the building;

c) allocating key logistic facilities and the site access along the site periphery; and

d) allocating the remaining facilities.

4   Principles in Allocating Temporary Facilities

The allocation of temporary facilities is complex due to the uniqueness of site conditions. The general principles mentioned above suggest a systematic approach for site planners to analyze and identify the SPLP sequence. The principles involved in allocating space for temporary facilities are discussed in the next section with an illustrated example.

5   Example of Site Production Layout Planning

The example shown in Figure 9.1 is a building project consisting of two 40-storey residential public housing blocks. The site space available for site facilities is moderately sufficient when compared with building developments in the private sector in Hong Kong. It is the contract requirements as stipulated by the Hong Kong Housing Authority that contractors should use mechanised methods and precast façade in constructing the concrete frame. The public housing blocks are basically of cross wall construction. The building consists of a cruciform with four wings: two long wings (A and C) and two short wings (B and D) opposite each other. Precast concrete façades are installed as the envelope for the residential units. They are installed prior to the fixing of wall or column reinforcement for the load bearing cross walls. Steel wall formwork is used as shutters for concreting. Aluminium formwork is used for the construction of beams and slabs. The typical floor construction sequence is shown in Figure 9.2. This example focuses on the SPLP for the construction of the concrete frame, which lies in the critical path of the project.

In this example, 16 temporary facilities are selected and allocated during the frame construction period. The allocation process follows the general priority stated in Table 9.2. The general principles in allocating the position and space for the facilities are demonstrated in the following example.

5.1   Tower crane

Tower cranes have great manoeuvrability because of their circular and vertical movements. The coverage of the working area by a tower crane is governed by two factors: the working radius and the hoisting capacity. Working radius, also known as the "sweep", is the basic and critical factor in deciding the location of a tower crane. The working radius should cover all loading points and unloading points. The loading points are usually located outside the building at the ground floor level. Unloading points are the positions at which the load is to be discharged and are usually at the working levels.

The hoisting capacity of a tower crane is proportional to the working radius. At the maximum working radius, the tip of the tower crane, where the point of maximum bending moment is located, provides the least hoisting capacity. Planners should therefore consider the weight of the loads when selecting a tower crane.

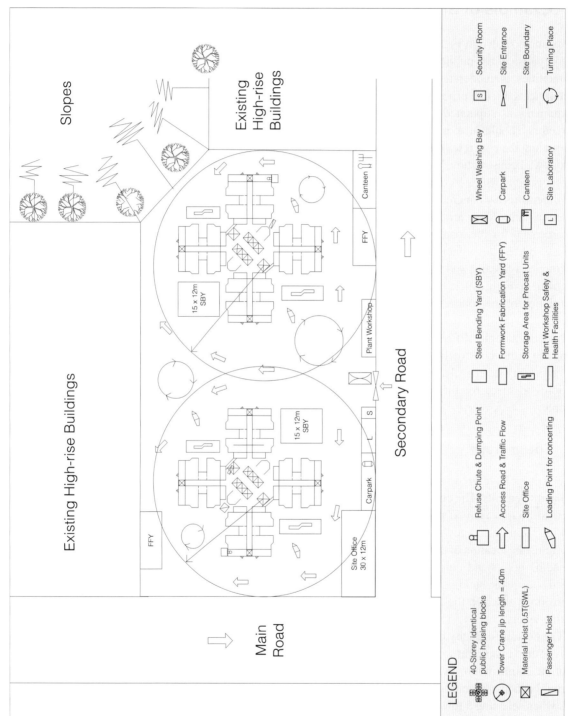

Figure 9.1   Site production layout plan for a public housing project

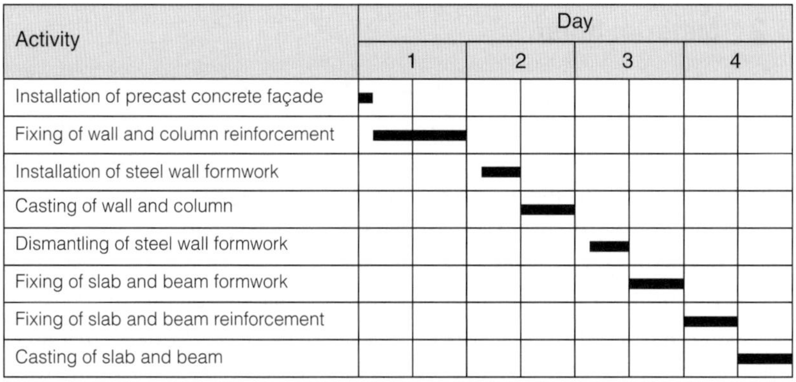

Activity	Day			
	1	2	3	4
Installation of precast concrete façade	▪			
Fixing of wall and column reinforcement	▬▬			
Installation of steel wall formwork		▬		
Casting of wall and column		▬		
Dismantling of steel wall formwork			▬	
Fixing of slab and beam formwork			▬	
Fixing of slab and beam reinforcement				▬
Casting of slab and beam				▬

Figure 9.2   A typical 4-day floor construction sequence for one wing

Besides the working radius, overlapping between two cranes is the second important factor to be considered. Overlapping frequently occurs when buildings are situated close to each other. If overlapping occurs, then the jibs should be set at different heights, as a general rule, of a minimum of 4 m to prevent crashing between the jibs. Planners should refer to the manufacturer's instructions regarding overlapping. In addition, the jib of a crane should not sweep over adjacent buildings or beyond the site boundary whenever possible.

The mast of a tower crane should be located where it will not interfere with construction activities. It is rarely located inside a lift shaft or machine room because this will delay lift installation. In most cases, a tower crane sits at a location where there is only relatively small quantities of non-critical work or at the light well. It is expensive to relocate a tower crane and work progress can be seriously affected.

In this project, the tower cranes are used for handling mainly precast façade, steel formwork, reinforcement bars and concreting. The precast concrete façades are the heaviest materials and can be up to five tones in weight for each panel. Although a tower crane of a smaller capacity can pick up the panel if it is stored closed to the centre of the crane, a crane with a maximum capacity of five tones at the tip offers more flexibility. In order to provide a good coverage of the fabrication yards, a tower crane with a 40 m-long jib is selected. Both tower cranes are situated at the light well at the centre of the buildings. The assigned sitting positions enable the tower cranes to cover the building blocks and a considerable amount of site storage areas around the building blocks.

5.2   Material hoist

The distance between a material hoist and the material storage areas is the major determinant in locating a material hoist. The material hoist should be close to the storage areas. On the other hand, it is important for lorries to have good access to the storage areas and the material hoist; otherwise, extra time will be required to transfer the materials between the storage areas and the material hoist.

A material hoist should be appropriately located to enable the delivery of materials to different work places over the shortest distance. Imbalances in delivery distance should be avoided so that travelling time can be saved. It is also advantageous to choose a location with the least effect on the building structure. Material hoists should therefore be installed against external walls where there are openings such as window openings or balconies, to act as access points for unloading materials. Forming and re-filling of temporary openings for access points can be very expensive and external wall finishing work can be delayed.

The material hoists for this project are located at the light well opposite to the tower cranes. This spreads the material storage areas evenly and thus avoids congestion upon material delivery.

5.3   Passenger hoist

Providing safe transportation for site personnel is the main factor influencing the allocation of a passenger hoist. For public housing projects, there is a term in the contract stating that the location of a passenger hoist should be agreed by the site supervision staff before installation, which requires the contractor to comply with all safety regulations covering the use of a passenger hoist. The provision of safe access is the most important factor. The passenger hoist should also be near the access road in order to shorten the walking time of workers.

An access landing is usually provided on the first floor and on every six floors thereafter as construction work progresses. Most passenger hoists are installed at window openings. The main benefit of this location is easy installation and dismantling. This is similar to the rationale behind the installation of a material hoist.

5.4   Refuse chute

The removal of refuse is important to the provision of tidy and clear working spaces for construction works. This is particular critical when finishing works have commenced since the discharge of refuse becomes a daily operation which should be carried out efficiently. Refuse chutes should be easily accessible for the disposal of rubbish at the working levels and the subsequent removal of the garbage tank at the collection point on the ground floor by refuse trucks.

Environmental consideration is another crucial factor in choosing the location of a refuse chute. Rubbish disposal produces a lot of noise and a lot of dust is also produced at the refuse chute outlet. To minimize noise and air pollution, a refuse chute should be located at the "dead corner" of the site where it is accessible but creates minimal nuisance to workers and adjacent properties.

In this project the refuse chutes are located at the "dead-quadrant" of the building blocks. Good access has been provided for disposal. Furthermore, the rubbish collection areas are paved to facilitate the discharge of surface water generated by the dust control water spraying system.

5.5   Access road

An access road is defined as a vehicular means for material transportation and plant movement within a construction site. When allocating space for access roads, two major factors should be considered. Firstly, the access roads should be able to reach all the material storage areas and loading and unloading points. Secondly, the road should be wide enough to minimize traffic congestion. If two-way traffic is not possible due to insufficient site space, passing place, lay-by or "turning place" should be provided for regulating the traffic when there is congestion.

In this project a six-metre wide access road is provided around the building. A turning place is located in front of the site entrance for regulating the traffic and queuing. Also, the access roads are widened at each block to provide passing places for traffic.

5.6   Site office

A site office should be located near the main site entrance for controlling access and to prevent vandalism. It should be at a position where it can command an overall view of the site. Offices should be large enough for site staff to carry out co-ordination work, conduct meetings and prepare documentation. The size of the site office is specified in some government contracts. Nowadays, 15-square metre containers are widely used as site offices because they are easy to erect and dismantle.

However, the working areas for sub-contractors are usually located at the working levels.

5.7 Loading point for concreting

Loading points are the positions where ready-mixed concrete is discharged from a truck into a skip for hoisting to the concreting position. It should therefore be within the sweep of the tower cranes. The location should be easily accessible by ready mixed concrete trucks, preferably with a turning place for discharging concrete into skips. The hoisting of concrete skips is one of the major hoisting activities; it will be safer and more efficient if loading points are within the crane operators' view. However, unlike the storage areas or the fabrication areas, it can be changed to suit the site conditions.

In this project, ready-mixed concrete is used because there is not enough space for installing a concrete batching plant. The loading points are located at the same quadrant where the tower crane is installed. However, additional loading points can be assigned at the opposite quadrant in order to shorten the hoisting times.

If a concrete batching plant is installed on a site, conrete trucks will also be used for transporting the concrete between the batching plant and the loading point since direct discharge of concrete from the mixer to the concrete skip is inconvenient and may cause a bottleneck. Concrete pumps can be used for concreting for such direct transportation.

5.8 Steel bending yard

The steel bending yard, especially the storage area for cut and bent steel bars, should be allocated within the working radius of the hoisting plant, such as mobile cranes or tower cranes. The transportation time can be reduced if they are closed to the centre of the tower crane.It should also be close to the access road for easy unloading of stock steel bars. The steel bending yards are located in the quadrants facing the turning place to facilitate the delivery of steel bars.

5.9 Formwork fabrication yard

A large space is required for the initial fabrication of formwork. As the formwork will be re-cycled during construction, a small amount of site space is asigned for formwork fabrication once the typical floor construction cycle has been established.

5.10 Storage area for precast units

Precast units are usually fabricated offsite and are delivered to site by truck on a daily basis to minimize the demand on site storage. However, it is important to consider that precast units demand a large space for storage. Also, they are heavy and it is a good idea for the storage area to be located close to the tower cranes and their permanent positions.

5.11 Plant workshop

A plant workshop should be located where it is easily accessible for the delivery of hand tools and mechanical parts. It is usually located next to the site office for easy delivery. Generally, a plant workshop is located at the site boundary to minimize the disruption to construction work.

5.12 Safety and health facilities

A canteen is usually provided if the site is situated in a remote area. The canteen should be located in a convenient and safe area for site personnel. It should also be at a position where it does not interfere with the construction activities.

Minimum requirements for toilets and washing facilities provided on site are specified in the government contract. Toilets should be located at a place that is convenient for workers. They should be installed not only on the ground floor but also on the upper levels at a reasonable interval; for example, on every four floors. Mobile toilets are used extensively on construction sites nowadays.

5.13 Facilities at site entrance

The location of the site entrance should be approved by the Highways Department. It is usually located at a secondary road where the traffic is light. The site laboratory is usually located at the site entrance for the convenience of taking samples from a ready-mixed concrete truck and delivery of samples for testing outside the site.

It is a statutory requirement that the wheels of trucks must be cleaned before they leave the site so that the road will not be contaminated by mud and dusty materials. Therefore, a wheel washing bay should be provided.

A carpark is usually provided when there is sufficient space . It should be convenient and safe for drivers to walk between the car park and site offices.

The example shown above discusses the general principles of SPLP. In the planning process, the allocation of facilities is governed by the complicated multi-interrelationship of planning constraints. This example demonstrates the basic strategies. More comprehensive analysis should be carried out for large and complicated projects.

6 Conclusion

Systematic research in SPLP is difficult due to the unique nature of site conditions on different building projects. The selection of temporary facilities, including production facilities, vehicular facilities and logistic facilities, depends on the construction schedule and method statements.

In high-rise building construction, materials handling is the key issue in optimizing SPLP with the aim of reducing transportation time. Tower cranes are therefore a critical plant in the planning process. Having allocated the sitting position of a tower crane, planners can determine other production facilities, site storage areas and fabrication yard around the tower crane. Other logistic facilities are usually located along the site boundary.

Site production planning is a complex task. There is hardly a standard or systematic approach for overall SPLP. This chapter discusses only the general principles that determine priority and the allocation of the facilities. More advanced study of techniques to optimize site layout planning is beyond the scope of this chapter.

References

1. Calvert R. E, Bailey G. and Coles D. 1995. *Introduction to Building Management*. 6th ed. Oxford: Butterworth Heinemann.

2. Cheng. M. Y. and O'Connor J. T. 1996. "ArcSite: Enhanced GIS for Construction Site Layout." *Journal of Construction Engineering and Management*. ASCE 122 (4): 329–336.

3. Leung, W. T. A, Tam, C. M. & T. K. L. Tong. 2001. "Decision Rules for site layout planning." *Proceedings of the First International Structural Engineering and Construction Conference*. 315–319.

4. Meyers Fred E. 1993. *Plant Layout and Material Handling*. Englewood Cliffs, N. J.: Regents/Prentice Hall.

5. Tommelein, I. D., Levitt, R. E., Hayes-Roth B and Confrey T. 1991. "SightPlan Experiments: Alternate Strategies for Site Layout Design." *Journal of Computing in Civil Engineering*. ASCE 5 (1): 42–61.

Chapter 10

Temporary Works in Building Construction

Hong-Xing WEN

Temporary works are necessary for the construction of any permanent structure. This chapter discusses issues immanent to temporary works for building construction, particularly in the context of Hong Kong. It firstly outlines the structural design principles and loads relating to temporary works. From the perspective of determining the striking time for concrete formwork, two issues are briefly discussed: the setting out of formwork striking criteria and the methods to estimate the properties of early-age concrete. Findings on the causes of temporary work related accidents and a case of collapse of building are then presented. Finally, legislation, codes of practice and guidelines relating to construction safety and temporary works in Hong Kong are outlined.

Temporary Works in Building Construction

1 Introduction

Temporary works may be defined as any structure or facility that provides temporary support for permanent structures during construction, or access or space to facilitate construction activities. There are many types of temporary works and their scale and complexity vary significantly. Examples of temporary works include notice boards, signposts, fencing, hoardings and protective walkways for pedestrians, shoring for excavation, support frames and foundations for tower cranes, access platforms and roads, concrete formwork and falsework, as well as access scaffolding.

Temporary works have a significant impact on the safety, quality, environment and profitability of a construction project. They normally account for 15–40% of the total cost of a project. For some civil engineering projects the percentage of cost can be up to 60% (Illingworth 1987). Effective cooperation amongst contractor, temporary works designers and permanent structural designers therefore makes economic sense.

This chapter discusses issues concerning temporary works in building construction, particularly in the context of Hong Kong. The following areas will be addressed in separate sections:

a) Design philosophy;
b) Loads on temporary works;
c) Methods for determining when to remove concrete formwork;
d) Safety;
e) Legislation, codes of practice and guidelines relating to construction site safety and temporary works in Hong Kong.

2   Design Philosophy

The design of temporary works is governed by the same principles that apply to the design of permanent structures, in terms of **safety**, **functionality**, and **cost**. Safety means the temporary works must have adequate strength and stability to resist expected loads. Functionality means the works must be sturdy enough to support construction without undue movement, excessive deflection or vibration which would hinder normal construction activities or result in a finished product of a substandard quality.

While structural design for permanent buildings takes into account the loading on the completed structure, design for temporary works must consider the whole lifecycle of the works, from scheme conception to final removal. An engineer must exercise considerable judgement in the design of temporary works and possess adequate knowledge of the following:

a)   construction process
b)   materials used
c)   techniques employed
d)   underlying principles and limitations of the design codes
e)   contractor's mode of operation
f)   structural mechanics
g)   geotechnics
h)   structural design principles
i)   available construction techniques and their relative costs

Many accidents occur because of inconsistency between design assumptions and actual conditions on site. In one such case, the engineer who designed the temporary works supporting a permanent concrete bridge deck never visited the site and was paid off before construction started. The contractor then erected and operated the temporary structures without being aware of the design assumptions and without inspection by the design engineer (Ratay 1987).

"Temporary" means the works will last for a limited period of time. Some design factors for permanent structures need to be adjusted when used for temporary works design. For example, the wind code of Hong Kong specifies wind load on permanent structures based on the strongest wind that has affected Hong Kong in the past 50 years. A temporary structure may last for only a few months or a couple of years at most. It would therefore be unduly conservative to adhere to the load specified for permanent structures. The code for falsework design, BS 5975, allows a reduction and sets upper limits for wind load.

The design strength of materials for permanent structures must take into account long-term effects such as creep and corrosion, which may not be a problem for temporary works. Certain design codes allow an increase in the design strength of materials for temporary works. (BS5268 structural use of Timber), for example, considers the load duration and allows an increase in the design strength for timber and timber products.

The majority of temporary works in Hong Kong are designed to British standards but other internationally recognized standards such as the ACI Committee 347 Report, Guide to Formwork for Concrete, are also accepted. Design codes for permanent structures are adopted for the design of certain temporary works with due consideration of loads and safety. Such design codes currently in use include BS5950 Structural Use of Steelwork in Building (limit state design), BS449 The Use of Structural Steelwork in Buildings (allowable stress design), and BS8110 Structural Use of Concrete, as well as internationally recognized design codes.

3 Load on Temporary Works

Temporary works have to cope with a more complicated combination of loads than permanent structures. The design of permanent structures has to take into account three main types of loads:

a) dead load from the structure's weight;
b) life load imposed by people, furniture, equipment and variable soil and hydrostatic pressure imposed on underground structures; and
c) wind load.

In addition to the above, the design of temporary works must consider other factors such as the construction operation load, the dynamic effects of construction equipment and concrete pressure. The actual load is influenced by the construction procedures adopted, so such procedures and site conditions must be known or clearly specified before the load can be determined.

A Guidance Note on Safety at Work (Falsework — Prevention of Collapse) issued by the Occupational Safety and Health Branch of the Labour Department of the Hong Kong SAR Government specifies loads under the following two categories:

Vertical loads:

a) The structure's own weight;
b) Permanent works to be supported;
c) Impact due to placing of permanent works (e.g. free fall of wet concrete);

    d) Construction operations (a minimum of 1.5 kN/m^2 should be allowed for the operations);

    e) Temporary storage of materials;

    f) Traffic load;

    g) Plant (the operating load should include the weight of plant, dynamic effects and vibration effects);

    h) Induced wind loads; and

    i) Uplift due to wind and flotation.

Lateral loads:

    a) Wind load;

    b) Hydrostatic pressure from wet concrete or an external source;

    c) Lateral earth pressure;

    d) Differential movements (e.g. ground movement) of supports;

    e) Vibration effects such as those arising from concreting or piling operations nearby;

    f) Flowing current;

    g) Unsymmetrical distribution of vertical loads, such as effects due to unbalanced concrete placement;

    h) Unsynchronised jacking of permanent works against falsework;

    i) Sway of falsework;

    j) Buckling of props;

    k) Eccentricity of vertical loads due to construction deviations, especially for falsework on sloping ground; and

    l) Dynamic effects of plant and equipment.

The guidance note also requires the designer to consider a minimum lateral load under the most adverse combination of the above mentioned lateral loads, or 2.5% of the vertical loads taken as acting at the points of contact between the vertical loads and supporting falsework, whichever is greater. BS5975 Code of Practice for Falsework takes a more conservative approach and recommends that falsework be designed for a horizontal force applied at the top of the falsework that is greater than all known horizontal load plus 1% of the vertical applied load or 2.5% of the vertical load.

    The minimum allowance of 1.5 kN/m^2 for vertical load imposed by construction operations is relatively small compared to the typical load imposed on a permanent structure. It represents the weight of the operatives placing the concrete or steelwork and associated light equipment, or the load imposed by a 60mm thick layer of concrete. (The load imposed by an average concrete slab is about 25 kN/m^3.) It would therefore be a very dangerous practice to heap concrete on the formwork during construction, because it could significantly overload the formwork/falsework.

Wind load on permanent buildings in Hong Kong is specified in the Code of Practice on Wind Effects Hong Kong — 1983. The wind loads given in the code are based on peak gust velocity over a 50-year period. The wind pressure specified in the Hong Kong wind code is equivalent to a gust velocity of about 44m/s (150km per hour). Generally, in most cases, construction activities will stop when wind speed exceeds the working wind limit, which is about 18m/s (65 km per hour) as specified in BS5975. BS5975 imposes an upper limit on the maximum wind load for which a falsework must be designed. It also specifies wind load factors for soffit and parapet forms.

Most concrete formwork designers in Hong Kong adopt CIRIA Report 108, Concrete Pressure on Formwork, for determining concrete pressure on formwork. In the method recommended by the report, the distribution of concrete pressure may either be triangular in shape, as derived from the principle of hydrostatic pressure shown in Figure 10.1a; or trapezoidal in shape as shown in Figure 10.1b, depending on a number of factors such as the dimensions of the member's cross section, the height of the member to be cast, the rate at which the concrete is poured into the formwork (or the rate at which the placed concrete rises), type of cement used, type of admixtures in the concrete, the concrete's workability, and temperature. When the member to be cast is short and the rate at which the placed concrete rises is relatively high, the concrete would be similar to a thick liquid during vibration compaction. In this case the pressure of the concrete is close to the hydrostatic pressure and is determined by a simple formula as shown below:

$$p = wh$$

Where p is the pressure of the concrete in N/m^2, w is the weight of the fresh concrete per unit volume in N/m^3, and h is the distance as measured from the top of the concrete in metres. For standard concrete, w is usually taken as 25 kN/m^3. The pressure envelope is triangular in shape as shown in Figure 10.1a.

When the member to be cast is tall and the rate at which the concrete is placed is relatively slow, the pressure on the lower part of the formwork initially increases with the increasing height of the concrete as in the case of a shallow formwork. As the height of the fresh concrete continues to rise, disturbance of the concrete at the bottom caused by compaction on top becomes less. The arching effect of the aggregates prevails and the pressure on the lower portion of the formwork eventually stops increasing. CIRIA report 108 recommends a **maximum design pressure**, p_{max}, for such cases, assuming the pressure of fresh concrete on the top portion of the formwork is equal to the hydrostatic pressure up to p_{max}, and that the pressure becomes constant down to the bottom of the formwork (Figure 10.1b).

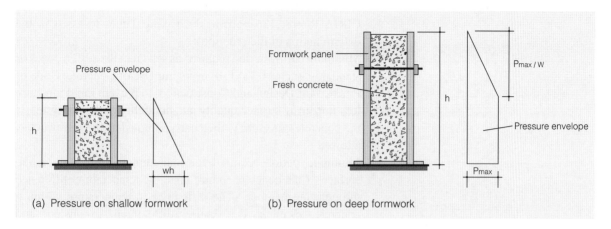

Figure 10.1   *Pressure of fresh concrete on formwork.*

4   Determining When to Remove Formwork

It is necessary to determine the right time for removing the formwork in order to facilitate construction and avoid collapse of or damage to the newly constructed structure. Many accidents occur each year due to early removal of formwork or falsework. The collapse of a high-rise building at Bailey's Crossroads in Fairfax County, Virginia, in the United States is due to the early removal of supporting props (details to be discussed later). Removing formwork early may cause shrinkage cracks to appear on the surface of the concrete and impair the durability of concrete members. On the other hand, delayed removal may result in damage to certain types of formwork and make it more difficult to repair visible defects.

The right time for removing formwork is determined by a set of **criteria** governing the curing of the concrete and the **time** needed for the concrete to meet those criteria.

The criteria include:

a)   Minimum strength required to prevent collapse/permanent damage of the reinforced concrete member through excessive stress on the concrete and the interface between concrete and reinforcement;

b)   Minimum strength (or modulus of deformation) required to prevent excessive deflection;

c)   Minimum strength or minimum time required to prevent surface concrete damage. (This criterion depends mainly on the type of formwork face material used and the quality requirement of the finished surface. Charts are available in technical reports such as

CIRIA Report 136 Formwork Striking Times to assist in the determination of the criterion);

d) Minimum curing period required as set out in the design codes or standards, such as BS8110 Structural Use of Concrete: Code of Practice for Design and Construction, to prevent rapid moisture loss impairing surface quality and durability of the concrete. (If the concrete cures early, then early removal of the formwork is allowed);

e) Minimum period required for maintaining insulation on the surface of the concrete to minimize the temperature difference through the member. (The difference in temperature within a concrete member can cause differential expansion and subsequent cracking (referred to as thermal cracking) in the concrete, impairing its strength and durability);

f) Other considerations, such as preventing freezing and thaw damage in cold weather concreting.

Before removing the formwork, one must estimate the properties of the concrete against the set criteria, which are often expressed in terms of minimum strength requirements. There are numerous methods for estimating the strength of in-situ concrete and they may be classified into two categories: one which relies on **direct measurement of the mechanical properties** of the concrete; and another which relies on **measuring the temperature** of the concrete and inferring the strength through maturity laws.

Methods relying on direct measurement of the mechanical properties include the following:

a) **Cubes stored alongside the concrete**

Cubes are cast using the same concrete as that in the structural members and stored alongside the member, so that the environmental conditions for the cube are fairly similar to those affecting the structure. The temperature inside the structure, however, could be much higher than the temperature in the cubes because of volume difference. The greater the mass, the greater the rise in temperature and the faster the strength increase. Using cubes, therefore, tend to underestimate the concrete strength inside a structure.

b) **Temperature-matched cubes**

Figure 10.2 illustrates the principle of the temperature-matched cube method. The cubes cast from the same concrete are stored in water after final curing. A thermometer is inserted into a pre-selected position in the newly cast structure and connected to

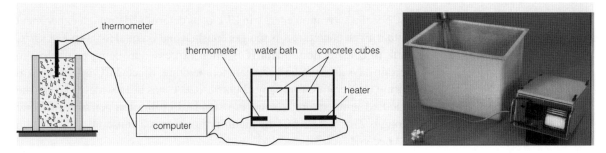

Figure 10.2   Illustration of temperature-matched cube test

Figure 10.3   Temperature-matched curing kit (Photo courtesy of Wexham Development Ltd.)

a computer. Another thermometer is inserted into the water and also connected to the computer. The computer compares the temperature inside the structure with the temperature in the water. If the water temperature is lower than the temperature inside the structure, the computer switches on the heater in the water until the two temperatures are equal. Since the test cubes and the concrete in the structure experience similar temperature history, the cubes in the water reflect more closely the actual strength of the structure than the cubes stored alongside the concrete members. Figure 10.3 is a photo of a temperature-matched cube kit.

c)   Other methods include penetration tests, pull-out tests and break-off tests, which are not widely used in Hong Kong. For more details the reader may refer to CIRIA Report 136. Sometimes the **rebound hammer test** is used for estimating concrete strength. It is non-destructive and easy to operate, but requires careful calibration.

The second category of methods for estimating concrete strength is based on maturity laws for concrete, which states that concrete from the same mix at the same maturity should have approximately the same strength, irrespective of the actual temperatures experienced.

Traditionally, maturity is defined as the sum of the time interval and the average temperature of the concrete during that time interval (Neville 1996). In recent years, the Arrhenius' equation for the rate of chemical reaction has been widely adopted for determining the maturity of concrete (Wen 1989). The maturity is determined by the following equation:

Equation (1)
$$M = \sum m \Delta t = \sum e^{-\frac{E}{R}\frac{1}{T+273}} \Delta t$$

In equation (1), Δt is the time interval and m is the maturity function. T is the average temperature, in centigrade, of the concrete during the period

Δt. E is the activation energy in kJ/mol as determined through experiment with a given concrete. R is the gas constant and is equal to 0.008314 kJ/°C mol. In construction, maturity is often expressed as being equivalent to certain days at a standard constant temperature. Since 20°C is the standard curing temperature for many countries, 20°C has been widely used as the standard temperature. (Note: the standard curing temperature in Hong Kong is 27°C). At 20°C, the maturity function is a constant, and equation (1) becomes

Equation (2)

$$M = \sum e^{-\frac{E}{R}\frac{1}{293}}\Delta t = e^{-\frac{E}{R}\frac{1}{293}} t_{20}$$

Letting equations (1) and (2) equal, the formula for the total elapsed time at a temperature of 20°C is obtained as Equation (3).

Equation (3)

$$t_{20} = \sum e^{\frac{E}{R}\left(\frac{1}{193} - \frac{1}{T+273}\right)}\Delta t$$

In equation (3), *t20* is termed the equivalent time at 20°C.

According to maturity law, the concrete strength should be a function of the maturity regardless of the temperature history. Research has proved that the maturity law is valid for most of the concrete within a certain period of time, particularly in the early stage of the concrete. There have been debates about the validity of the maturity function for concrete containing pozzolans such as PFA (pulverised fuel ash), silica fume and GGBS (Ground Granulated Blast-furnace Slag). This is mainly because the reaction of pozzolan in the concrete is a "secondary" reaction which follows cement hydration. Nevertheless, research has shown that maturity law is valid for young concrete containing pozzolans (CIRIA 136, Magne 1986). Figure 10.4a shows the data obtained from tests on high strength concrete cubes containing silica fume, subjected to variable temperature histories (Wen 1989). The temperature histories of the tested cubes are illustrated in Figure 10.4b, in which ADI means adiabatic condition and TPS means temperature profile simulation. Figure 10.4a indicates that the maturity law is valid for the concrete up to about seven days.

Table 10.1 is an illustrated example of the calculation of equivalent time at 20°C for a newly cast concrete. The average temperature of the concrete is obtained over four-hour time intervals. The maturity function used in this calculation is Equation (3) with the activation energy being 38.5kJ/mol (Wen 1989). At the end of the day, the equivalent time at 20°C is 79.28 hours or 3.3 days.

Figure 10.5 is a typical concrete strength-equivalent time relation chart. If, for example, the minimum required compressive strength for stripping of the formwork is 20N/mm², the minimum equivalent time at

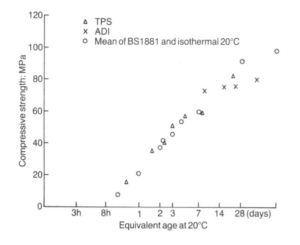

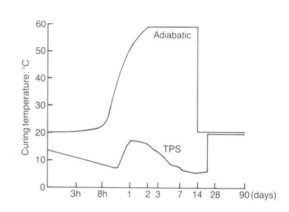

(a) Compressive strength vs. equivalent age relation for a concrete containing silica fume subject to variable temperature histories

(b) Temperature history profiles of the concrete shown in (a)

Figure 10.4 Strength of concrete containing silica fume subject to variable temperature plotted against equivalent time at 20°C

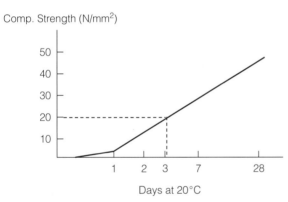

Figure 10.5 A typical strength equivalent time relation for concrete

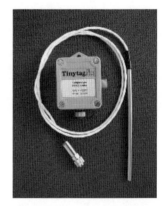

Figure 10.6 Maturity meter (photograph courtesy of Wexham Development Ltd.)

Elapsed time	0-4hrs	4-8hrs	8-12hrs	12-16hrs	16-20hrs	20-24hrs
Average temperature	25°C	25°C	30°C	45°C	60°C	55°C
Time increment Δt	4 hrs	4 hrs	4 hrs	4 hrs	4 hrs	4 hrs
Maturity function*	1.3	1.3	1.68	3.46	6.68	5.4
Maturity function Δt	5.2hrs	5.2hrs	6.72hrs	13.84hrs	26.72hrs	21.6hrs
Total equivalent hours at 20°C = 5.2+5.2+6.72+13.84+26.72+21.6 = 79.28hrs						

* The maturity function used for the above calculation is Equation (3)

Table 10.1 An illustrated example for the calculation of equivalent time at 20°C

20°C for removing the formwork is just over three days. In the example presented in Table 10.1, the concrete would satisfy the formwork removal criterion after one day.

The strength-maturity (or equivalent time) relation is job-specific, i.e. the relation is unique for a given concrete. Therefore careful calibration of the relation is required for each type of concrete. Some concrete suppliers provide such information. Research has shown that methods based on the maturity law generally yield very close prediction, provided the strength-maturity relation is accurate.

Instruments are available in the market for determining the maturity of in-situ concrete like the one shown in Figure 10.6. To determine the maturity of concrete, a temperature probe is inserted into the concrete to measure the temperature. The maturity is automatically calculated according to a chosen maturity function.

From the above discussion, the procedure for the determination of formwork removal time is as follows:

a) Determine formwork removal criteria, which can be specified in terms of strength or minimum time as set down in design specifications or standards such as BS8110.

b) Estimate the properties of the concrete. This can be done by testing sample cubes or other methods of determining the mechanical properties of the concrete in the structure, or from maturity laws with the assistance of a maturity meter.

5 Safety Related to Temporary Works

Building construction has always been a high-risk industry. In 1997, about 40% of the total industrial accidents in Hong Kong is construction industry-related. The industry's accident rate in Hong Kong in the past ten years ranges from 200 to 470 per thousand workers per year. There are dozens of fatal construction-related accidents each year, many of them associated with handling of or working on temporary works (Mak, 1998).

Research and investigations around the world indicate that accidents arising from the failure of temporary works often have their root in management deficiencies, although technical reasons were the direct and apparent cause (Ratay 1987). Technical reasons refer to shortcomings in the design, erection and dismantling of temporary works. Management deficiencies include failure to implement proper site safety procedures, insufficient inspection, inadequate communication between temporary works designers and erectors, ambiguity of responsibilities among architects, engineers and contractors, and between main contractors and sub-contractors.

A fatal accident caused by the collapse of temporary works is often triggered by apparently a trivial mistake, such as the accidental removal of a prop, a missing coupler on a scaffold, or a loose wedge on a sloping falsework. One example was the collapse of the Skyline Plaza at Bailey's crossroad in Fairfax County, Virginia, in the United States in 1973 (Philpott, 1987). The accident is illustrated in Figure 10.7.

The building was a reinforced concrete residential apartment with a regular floor plan through its height. The building was built at a cycle of seven days a storey at the beginning of the construction up to the 23rd floor, Figure 10.7a. The construction was then accelerated by working on 24-hour shifts and the cycle was reduced to five days per floor, (Figure 10.7b). After the 24th floor slab was cast, the back prop beneath the 23rd floor slab was removed, when the floor slab was only 10 days old, (Figure 10.7c). The top floor slab collapsed and the failure triggered progressive collapse of the lower floors through the height of the building, (Figures 10.7d, 10.7e, 10.7f). When the collapse occurred other trades were working on the lower floors and the tragic accident killed 14 workers and injured 53.

Back props provide the support for floor slabs after the formwork has been removed and are needed for most multi-storey reinforced concrete buildings. The load of the floor slab of such buildings is typically around 4.5–7.5 kN/m^2 (200–300mm thick slab) and the typical imposed load is around 2.5–4.0 kN/m^2. Thus, about two-thirds of the capacity of a floor slab is for carrying its own weight and only one third is for carrying imposed life load. Removing the back props when the top slab is not self-supporting forces the slab underneath to carry the weight of two slabs plus falsework and construction loads. In such a scenario, the supporting slab would almost certainly collapse (Figure 10.8).

Back propping for high-rise reinforced concrete buildings in Hong Kong is a common practice with one-level and two-level propping being the most common (Figure 10.9). The choice of the type of propping depends on the design loads on permanent structure, dimensions of the building plan, the sequence and manner in which back props are installed and the speed of construction. Many simplified methods for the determination of load shearing among props and slabs have been proposed. Some of these are included in design codes. Software for more accurate computer analysis is also available.

Some temporary works are vulnerable to bad workmanship. A loose joint or a missing wedge on sloping foundation could trigger catastrophic accidents, as Figure 10.10 illustrates. Figure 10.10a shows a steel tube

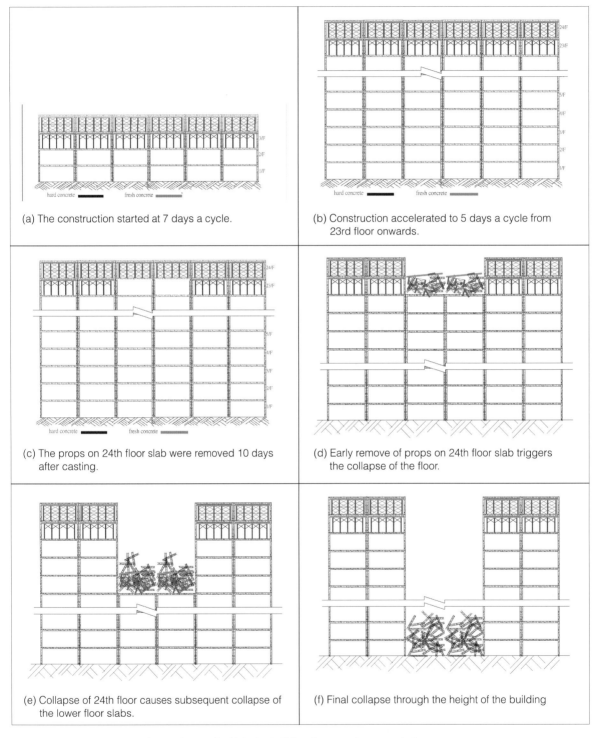

(a) The construction started at 7 days a cycle.

(b) Construction accelerated to 5 days a cycle from 23rd floor onwards.

(c) The props on 24th floor slab were removed 10 days after casting.

(d) Early remove of props on 24th floor slab triggers the collapse of the floor.

(e) Collapse of 24th floor causes subsequent collapse of the lower floor slabs.

(f) Final collapse through the height of the building

Figure 10.7   An illustration of the collapse of a high-rise building due to early removal of props

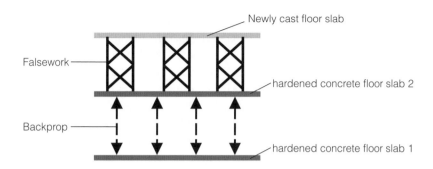

Figure 10.8   The early removal of back props before the newly cast top slab can be self-supporting would cause overloading of slab 2

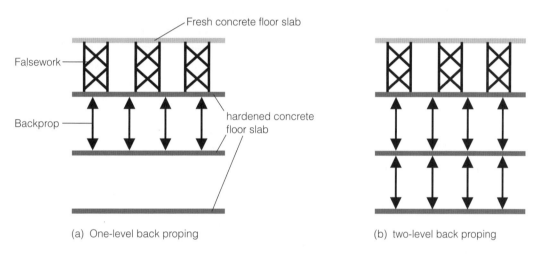

(a) One-level back proping

(b) two-level back proping

Figure 10.9   Illustration of one- and two-level backprops

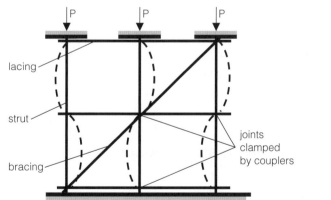

Figure 10.10a   Possible buckling mode and load capacity of properly erected falsework (all joints are clamped by couplers)

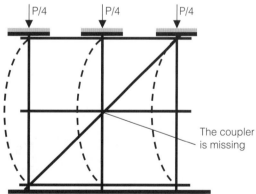

Figure 10.10b   Possible buckling mode and load capacity of the falsework when the central coupler is missing

falsework with vertical struts, lacings (horizontal members) and bracings (diagonal members) clamped firmly at *every* joint by couplers to form a stable structure. The props, if loaded to its capacity, is likely to buckle as indicated by the dotted lines in the figure. If, as shown in Figure 10.10b, the coupler in the centre joint is carelessly omitted or loosely installed, the effective length of the vertical struts is doubled and the capacity of the props would be reduced by approximately four times. Proper site inspection should be able to identify such pitfalls and minimize the probability of accidents.

Hadipriono et al. (1986) investigated 85 falsework collapses on bridges and buildings and identified **triggering**, **enabling** and **procedural** causes for failures. Triggering causes are those not directly related to design, erection, and dismantling of the falsework but associated with external factors or improper operation. These include improper falsework removal, the impact of construction equipment and vehicles, and the impact of concrete falling from height. Enabling causes are those more directly associated with the design and construction of the falsework and include inadequate lacing and bracing; inadequate number of props and shoring; and improper installation and maintenance of falsework and equipment. Procedural causes include inadequate review of falsework design and construction, lack of inspection of falsework and formwork during concreting, employment of inexperienced or inadequately trained workmen, and change of temporary works design without the consent of competent engineers or without proper redesign.

Investigations into construction accidents indicate that failure to appreciate the seriousness of accidents and reluctance to implement proper safety supervision due to contractors' obsession with profit is often the underlying reason for accidents. In recent years, new safety management systems for construction have been proposed and enforced through legislation in many countries.

The Management of Health and Safety at Work Regulations 1992 (United Kingdom), for example, requires contractors to prepare risk assessments to identify potential risks and ensure that the seriousness of the risk is properly evaluated by the contractor.

6 Safety Legislation and Guidelines in Hong Kong

In Hong Kong, the legal requirements for construction safety are set out in the Factories and Industrial Undertakings Ordinance (Cap 59), the Occupational Safety & Health Ordinance (Cap 509), the Buildings

Ordinance (Cap 123) and their subsidiary regulations, such as the Construction Site (Safety) Regulations (Cap 59I). Guidelines on temporary works and site safety have also been published by relevant government departments and authorities. These include Guidance Notes on Safety at Work (Falsework — Prevention of Collapse) published by the Labour Department; and the Code of Practice for Site Safety Supervision published by the Buildings Department.

The Factories and Industrial Undertakings Ordinance (Cap 59) and Occupational Safety & Health Ordinance (Cap 509) specify the responsibility and legal liability of the main and sub-contractors. The Construction Sites (Safety) Regulations (Cap 59I) sets out the legal requirements for temporary works such as crane, lifting gears, hoists, scaffolds, working platforms and excavation works.

The **Supervision Plan**, which was introduced into Section 2 (1) of the Buildings Ordinance (Cap 123) in 1996, sets out the safety management plan for building works or street works to be lodged by an authorized person registered with the Building Authority prior to or at the time of application for consent to commence building works or street works.

The **Technical Memorandum for Supervision Plans** and **Draft Code of Practice for Site Safety Supervision** were published in 1997 by the Buildings Department under Section 39A of the Buildings Ordinance. A corresponding supervision system was implemented for projects under the Buildings Department's control. The Technical Memorandum sets out principles and scope of the supervision system, including the principles for the plan's preparation, form and content. The Code of Practice was a more comprehensive document providing detailed guidelines for implementing the system.

A supervision plan comprises an **outline plan**, which sets out the site safety management structure; and a **detailed plan**, which includes specific safety requirements for certain construction works. Site safety management normally comprises three functional streams led by the architect, the structural engineer and the contractor respectively. Under each stream, **technically competent persons** (TCPs) are appointed to carry out routine site supervision. There are five grades of TCPs with T1 being the lowest and T5 the highest. Grades T1 to T3 are appointed for **routine site safety supervision** and T4 and T5 for **engineering supervision**. The qualification and experience requirements for TCPs are specified in the code of practice to ensure that the TCPs are competent for the tasks assigned to them.

The Code of Practice was refined in November 2000 to specify more clearly the division of responsibilities for temporary works amongst the three functional streams.

For the purpose of **temporary works supervision**, responsibilities are grouped into three cases:

a) **Case 1** includes those temporary works stipulated in the supervision plan provided by the Buildings Authority. The architect, the structural engineer and the contractor have their own responsibility in accordance with the approved plans.

b) **Case 2** includes works not required to be stipulated in the supervision plan and have no overstressing or overloading effects on permanent structures. The contractor has the sole responsibility for the design, construction and safety of these works.

c) **Case 3** includes works not required to be stipulated in the supervision plan but may cause overstressing or overloading effects on permanent structures. The contractor is responsible for the design, construction and safety of these works and must submit details related to the temporary works to the structural engineer for approval.

The Supervision Plan addresses three important issues underlying site safety for temporary works:

a) It encourages serious **implementation** of safety plans by establishing a clear safety management structure on site and specifying the frequency of site inspection required;

b) It requires the appointment of **TCPs** on construction sites and, by specifying the academic qualifications and experience required for different grades of TCPs, eliminates possibilities of design faults, as well as improper construction and inspection of temporary works by incompetent persons;

c) It specifies the class of supervision required for building works according to their complexity and the risks involved, hence encouraging effective use of resources.

7 Conclusion

The design and handling of temporary works affects the quality of permanent building and the safety of workers and the public. Since temporary works take up a significant portion of project cost, careful choice of temporary works makes a lot of economic sense.

An engineer designing temporary works must be familiar with the construction procedures and exercise considerable engineering judgement. Design assumptions based on the sequence of construction, the materials used and the load allowed for should be clearly specified and presented with drawings of temporary works to avoid ambiguity or misunderstanding between the contractor and the designer. Inspection of temporary works must be carried out by experienced competent persons who understand the design principles.

Determining when to remove formwork and estimating the properties of young concrete are important to ensure the good quality of the concrete, the smooth running of the construction operation, and the economic use of formwork, falsework and human resources. Theoretical methods could be adopted to assist physical testing in estimating concrete properties to save cost or improve reliability.

The majority of construction accidents are associated with temporary works. Causes include improper construction, inadequate inspection, design faults and lack of communication among the parties involved. Defective management systems are often the underlying cause.

Control of construction safety in Hong Kong has been tightened in recent years, as evidenced by the introduction of the Supervision Plan, which requires the implementation of site safety management and the appointment of technically competent persons for site and temporary works inspections.

References

1. Hadipriono, et al. 1986. "Analysis of Causes of Falsework Failures in Concrete Structures." *Journal of Construction Engineering and Management* Vol. 112, No. 1: 112–121.

2. Illingworth, J. R. 1987. *Temporary works — their role in construction.* London: Thomas Telford.

3. Mak, H. K. D. 1998. "Legislative control regime for ensuring safe use of scaffolds." Symposium on Bamboo and Metal Scaffolding. Hong Kong. 23 Oct 1998. 23–33.

4. Magne, Maage 1986. "Strength and heat development in concrete: influence of fly ash and condensed silica fume." ACI SP 91 Vol. 2: 923–940.

5. Neville, A. M. et al. 1994. *Concrete Technology.* Longman Scientific Technical.

6. Philpott, D. H. 1987. "Management of Temporary Works in the UK." Temporary Structures in Construction Operations, Proceedings of a session in conjunction with the ASCE Convention. Atlantic City, New Jersey. 29 April 1987. Ed. Ratay R. T. 48–56.

7. Ratay, R. T. 1987. "Temporary Structures in Construction Operations." Proceedings of a session sponsored by Construction Division of the American Society of Civil Engineers in conjunction with ASCE Convention in Atlantic City, New Jersey. 29 April 1987.

8. _____. 1996. *Handbook of temporary structures in concrete*. 2nd edition. New York: McGraw-Hill.

9. Wen, H. X. et al. Dec 1989. "The strength development of silica fume concrete and its prediction under varying temperature conditions." *Magazine of Concrete Research*. 41 (149): 199–204.

Chapter 11

Conditions and Constraints Governing the Use of Formwork Systems for Complex High-rise Buildings in Hong Kong

Raymond W. M. WONG

Nowadays, most clients require projects to be completed in the shortest time possible as a means of minimizing costs. For high-rise buildings, the most effective way to speed up work is to achieve a very short floor cycle, i.e. to have the structure of a typical floor completed within the shortest time. Aiming purely at speed often compromises the achievement of other quality targets. Problems such as misalignment, misplacement, defective concrete, or delay of other works causing serious interruption may result. This chapter, supplemented with several recent Hong Kong case studies, highlights the conditions and constraints governing the use of formwork and illustrates the practices and methods used by the local construction industry to construct complex buildings of various kinds.

11

Conditions and Constraints Governing the Use of Formwork Systems for Complex High-rise Buildings in Hong Kong

1 Introduction

Formwork systems are among the key factors determining the success of a construction project in terms of speed, quality, cost and safety of works. Nowadays, most projects are required by the client to be completed in the shortest time possible to minimize costs. For high-rise buildings, the most effective way to speed up work is to achieve a very short floor cycle, i.e. to have the structure of a typical floor completed in the shortest time. This can be done in Hong Kong within three to four days' time, even for buildings with floor areas up to 2,500 sq m. The key to achieve this, again from the production point of view, is by the use of a system of efficient and appropriately designed formwork. Modern buildings can be very complex in terms of scale, architectural or structural disign, and sophisticated building services or facility requirements. The design and use of the right formwork system, as well as an effective resource planning strategy to control and maximize the use of the formwork, is crucial to the overall success of a project.

Aiming purely at speed often sacrifices the achievement of other quality targets. Problems such as misalignment, misplacement, deflective concrete or holding up of other works causing serious interruption may result.

With reference to serveral recent Hong Kong case studies, this chapter highlights the conditions and constraints in the application of formwork and illustrates the practices and methods that the local construction industry uses to construct complex buildings of various kinds.

2   Classification of Formwork

Formwork can be classified according to a variety of categories related to the differences in size, location of use, construction materials, nature of operation, or simply the brand name of the products.

2.1   Classification according to size

Classification according to the size of formwork can be very straightforward. In practice, there are only two sizes: small and large. Any formwork that is designed for manual operation by workers is small. Very often, an erection process to be handled by a single worker is preferred, with site work best done independently to avoid possible waiting times. Due to reasons of size and weight, the materials and construction of small-sized formwork are thus limited. At present, the most common systems are made of timber and aluminium, and they are usually in the form of small panels.

In cases where large-sized formwork is used, the form is designed to be as large as practicable to reduce the amount of jointing and minimize the amount of lift. The stiffness required by large-sized formwork can be dealt with by the introduction of more stiffening components such as studs and soldiers. The increase in the weight of the formwork panels is insignificant as a crane is used in most cases.

2.2   Classification according to the location of use

Different elements in the structure of a building have different design and performance requirements in terms of the formwork used. A number of formwork systems are particularly designed for constructing internal or external walls, vertical shafts, columns, beams, and floor slabs. However, there are not many effective formwork systems for stairs and staircases. The complicated three-dimensional nature of an element involving suspended panels and riser boards as well as the need to cope with very different spatial and dimensional variances as required by individual design situations cannot be handled by a universally adaptable formwork system (Figure 11.1).

Figure 11.1   Typical detail of constructing a staircase using traditional timber formwork

2.3   Classification according to materials of construction

The types of materials used for formwork are traditionally quite limited due to the need to strike a difficult balance between cost and performance. Timber in general is still the most popular formwork material for its relatively low initial cost and adaptability (Figure 11.2). Steel, in the form of either hot-rolled or cold-formed sections and in combination with other sheeting materials, is another popular choice. In the past two to three years, full aluminium formwork systems have been used in some cases but the performance is still being questioned by many users, especially in consideration of cost and labour control (Figure 11.3). Other types of metals and alloys are still uncommon on construction sites due to their cost and easy substitution by other common metals.

The huge amount of tropical wood being used for formwork each year has resulted in criticism from environmentalists, while the continual escalation in timber prices has raised concerns about cost. As a result, there has been a strong trend towards the use of other formwork materials or systems to replace timber. At present, steel forms are gaining in popularity in Hong Kong, especially with the incorporation of semifabricated construction techniques under design-and-built arrangements (Figure 11.4).

Figure 11.2   Traditional timber formwork system is still very popular for large-size projects due to its adaptability

2.4   Classification according to nature of operation

Formwork can be operated manually or by other power-lifted methods. Some systems are equipped

Figure 11.3   Example of using aluminium formwork in a recent residential construction project

Figure 11.4   A government project involving the use of innovative precast elements in the construction process

with a certain degree of mobility to facilitate the erection and striking processes, or to allow horizontal movement using rollers, rails or tracks.

Timber and aluminium forms are the only manually-operable types of formwork. They are designed and constructed to facilitate independent handling without the aid of any lifting appliances. They are labour-intensive and are more appropriate for use on simpler jobs. However, such systems are also used in very large buildings with a big horizontal spread that feature complicated layouts so that the flexible nature of manually operated formwork can be made use of fully.

Power-lifted formwork can be of the self-climbing or crane-lifted types. Self-climbing formwork uses built-in hydraulic (Figure 11.5) or screw jack systems under a full-form or sectioned arrangement. Since the lifting power of the jacks is enormous, a supporting gantry system for the erection of the formwork panels as well as an enclosed scaffold system with inner and outer work platforms are usually provided to form a convenient and self-supporting work station for casting works.

Crane-lifted systems are usually in the form of large panels (sometimes called the gang form). They are fabricated either in steel sections and sheeting, or using plywood sheeting stiffened by metal studs and soldiers. These large panels can be stood on solid slab or fixed on brackets in case they are used for external walls or shafts (Figure 11.6).

Figure 11.5   Example of a self-climbing formwork with built-in hydraulic system to assist in the lifting operation

Figure 11.6   Example of a steel panel form for the casting of an independent-standing wall

Figure 11.7   Example of gantry-type formwork system specially designed to construct identical and repeated sections

Other forms include the gantry-type, travelling or tunnel form. These are more suitable for use in long repeated sections such as in railway stations, terminal buildings or other large horizontal structures. Recent Hong Kong applications include the construction of the 600 metre-long elevated expressway for the Lantau Link (Figure 11.7), the Airport Railway's Ground Transportation Centre (Figure 11.8), and West Rail Siu Hong Station (Figure 11.9).

Figure 11.8   *Another example of gantry form suitable for the forming of complicated shaped roof structure*

Figure 11.9   *KCR West Rail Siu Hong Station — gantry form enables the structure to be supported on piers and work on an elevated level on top of a flood discharge channel*

2.5   Classification according to brand name

Several patented or branded formwork systems have successfully entered the local construction market in the past decade. These include products from SGB, RMD, VSL, MIVAN, Thyssen and Cantilever. Each of these firms offers its own specialized products. Some can even provide a very wide range of services including design support or tender estimating advice. As the use of innovative building methods gains more attention, advanced formwork systems will become more widely adopted. The input of the well-established formwork manufacturers is no doubt a contributing factor.

3   Technical Considerations When Using Formwork

The selection and application of formwork, particularly for large-scaled and complex projects, depends on the following factors:

3.1   Design-related factors

The shape of the building

Simple block-shaped buildings are much easier to construct than buildings in awkward shapes, such as projects with curved, inclined, stepped, undefined or sculptured features. As a general rule, awkwardly shaped buildings can be more easily dealt with by using more traditional, labour-intensive formwork systems because of their better adaptability.

Design of the external wall

Some buildings may have many architectural features on the building exterior such as fins or ribs, sunshading blades, planter boxes, deep rebate windows or hoods for air-conditioner units. These may limit the choice of system-type formwork due to features that interrupt with the casting process.

Internal layout

Some buildings may have very simple layouts with few in-situ walls and floor plates framed with regularly spaced columns, as seen in many commercial and office buildings. However, some developments feature very complicated, load-bearing internal walls that can make the casting process difficult.

Structural forms

Like internal layouts, the structural form of buildings also affects formwork options. For example, buildings with a structural core in the form of a vertical shaft limit the use of other formwork systems other than those of a self-climbing nature. Buildings with flat slabs make table forms or flying forms the most viable choices. For buildings with regularly arranged shear wall designs, the best selection is large-panel type steel forms or other types of gang forms.

Consistency in building dimensions

Some buildings may have non-standardized dimensions due to the architectural design and layout or structural requirements. These include the regular reduction of sizes for beams, columns and walls in high-rise buildings as the structure ascends. Some formwork systems, like the climb form or steel form, may be quite difficult to use in such situations, for the frequent adjustments of the form to meet the changes in dimensions may eventually incur extra cost and time.

Headroom

Higher headroom increases the amount of falsework required and can also create accessibility and safety problems. It can also make the erection of formwork and the placing of concrete more difficult. Working

Figure 11.10   A transfer plate structure 20 m above ground level being cast at an advanced construction stage

headrooms of more than five metres are frequently found in buildings with transfer structures (Figure 11.10), entrance foyers, atriums inside shopping malls and many other functional, institutional and public buildings.

Building span

Large building spans also create problems similar to those with high headrooms. In addition, long-span structures generally have larger beam sections, heavier reinforcement provisions, or accompanying post-tension works. This will further complicate the formwork's design and erection process.

Repetitive nature

High-rise block-shaped structures usually require highly repetitive cycles and this is favourable to the use of formwork. However, the degree of repetition in buildings with a large footprint or underground structures such as basements is limited, and the use of formwork, as an expensive resource, becomes very critical in terms of cost and progress control.

Surfaces finishes

Fair-faced concrete demands very high quality formwork in terms of surface treatment of the panels, tightness of the formwork joints and dimensional accuracy. Requirements are slightly relaxed where the concrete surface is to be finished later.

3.2   Construction-related factors

Complexity of the built environment

Exceptionally small or large sites (Figure 11.11), sloped (Figure 11.12) or very crowded sites (Figure 11.13), proximity to sensitive structures, sites where other major activities are underway, or sites with many physical or contractual restrictions will increase the difficulty of working with formwork. These problems are tackled according to individual circumstances.

Speed of work

When working with buildings with large construction areas and horizontal spread, projects can be expedited by the introduction of additional sets of formwork to create more independent work fronts. This will, of course,

Figure 11.11   Extremely large-size and complex project —
the International Finance Centre Phase II

Figure 11.12   The Belcher's Garden — a 25,000 m²
commercial and residential mixed
development on a sloped site

increase the cost of production. For high-rise buildings, increasing the number of formwork used cannot always speed up the construction process and increase the speed of work, for the critical path still depends on the floor cycle. However, a properly selected, designed and arranged formwork system will increase work efficiency in each typical cycle. In some cases, adding a half or a full set of formwork, especially for floor forms, may help to speed up the cycle as the additional set can provide more flexibility when the form is struck at an earlier time.

Re-use of formwork

The re-use of traditional timber formwork is usually limited due to the durability of the plywood sheeting. The optimum number of times a timber form can be used is usually 12 to 14. Thus, it is still sufficiently economical to use timber formwork for high-rise buildings. Although metal forms can be reused more intensively, the high initial cost of providing the form often discourages its selection, especially when there is no need to recycle them too many times, for example in a low-rise development. A careful balance of cost, speed, performance, and the quality of output should be properly considered when making the selection.

Figure 11.13   An urban renewal
project in the congested
downtown environment
in Central, Hong Kong

Construction planning and arrangement

Construction planning such as the phasing or sectioning arrangement, integration of the structure, site layout and set-up arrangements or hoisting provisions and concrete placing facilities are important factors when considering formwork selection and application.

Area or volume of cast per pour

The optimum volume of cast per pour depends on the type of formwork used, the particular elements of structure to be placed, the actual scale of work, and different levels of provision of plant facilities. Usually a volume of concrete ranging from 60 to 200 cubic metres per pour can be comfortably handled in most site environments. It also depends on whether the concrete to be placed is for the vertical elements only or also includes the beams and slab, as a means of saving an additional phase in the overall work cycle.

Figure 11.14   A main beam with slot provisions to receive precast secondary beams in the KCRC Hung Hom Railway Station extension project

Involvement of other construction techniques

The application of tensioning and prefabrication techniques are often involved in the construction of high-rise buildings. This may have an impact on the use of formwork, especially where precast elements are to be incorporated during the casting process (Figure 11.14). Allowances should be made for additional provision of temporary supports or slot spaces and boxing out positions in the formwork for the precast elements, or extra working space for placing stressing tendons and onward jacking.

Dependence of work

Many factors should be considered before employing a construction plan and selecting the right formwork system. These include considerations of whether there will be lifting appliances provided for the erection of formwork; whether these appliances will be able to access the work spot to assist in the operation as the structural works proceed; whether any special equipment will be required for striking the forms; and how the removed formwork panels can be transported to other spots to continue working.

Figure 11.15   Typical sectioning arrangement in a building project incorporating complicated construction joints

Provision of construction joints

Sometimes a large number of construction joints is inevitable in a large structure because of the sub-division of works into effectively workable sizes. The provision of construction joints can challenge the output and affect the quality of the concrete (Figure 11.15). Careful selection should be made to ensure a particular formwork system can satisfactorily allow such arrangements.

Accessibility

During the course of construction, accessibility problems may be created through segregation, temporary discontinuation, or blocking of the layout by the partially completed building (Figure 11.16). Or, in cases where a shaft-type core wall is constructed in an advanced phase, the shaft may stand independently for a long period of time before it is connected to the horizontal elements. Arrangements for access to work places should be properly arranged when carrying out construction planning.

Figure 11.16   A large site with varied work frontages makes access dangerous and difficult

Feasibility of introducing alternative designs

Under the traditional separate design and construction procurement system, architects often design a building which is not suitable for the use of more advanced and efficient formwork systems. It is quite common for builders to submit alternative design proposals to clients for consideration with minor structural or architectural amendments so that more effective formwork methods can be applied. Very often, the cost benefit derived is shared between the builder and the client in order to achieve a win-win situation.

4   Examples of Application

The following are a number of recent construction case studies for illustrating the use of formwork systems on unique projects.

4.1   Festival Walk — using traditional manual-type timber formwork

The development is a shopping mall and leisure centre that comprises a four-level basement and a seven-storey upper structure. Built on a 21,000 sq m site, its design and construction features include:

a) a 48 m – span skating rink constructed of eight in-situ cast and post-tensioned beams, supported on the sides by bearers (Figure 11.17);

Figure 11.17   A 48 m – span transfer beam system in the Festival Walk provides structural support for the 4-storey office block

Figure 11.18   The forming of a circular vehicular
                ramp down into the basement
                using traditional timber formwork

Figure 11.19   Isolated island-type building portions formed
                during the construction stages by the atrium
                spaces that cut the site in between

Figure 11.20   Formwork detail for the general floor systems

Figure 11.21   Complicated work phasing and
                sectioning arrangement — a closer view of
                the gigantic structure

    b) a basement constructed using the top-down approach;

    c) a 40 m diameter circular ramp down into the basement also constructed in the top-down manner (Figure 11.18);

    d) a number of large span structures up to 32 m in length cast in situ, the majority of which were post-tensioned;

    e) 3 atrium spaces, averaging 35 m in span with 25 m headroom (Figure 11.19);

    f) an average headroom of about 4.5 m on each floor.

Due to the specific functions of the building, the layout of the 160,000 sq m building has few repeatable elements, making the application of system formwork unfeasible. As a result, traditional formwork with timber panels was employed throughout the project (Figure 11.20). Since

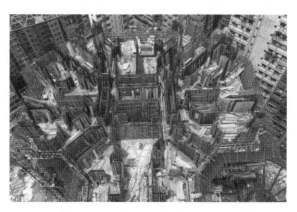

Figure 11.22   Podium structure with elevated portions extending from pier supports above formed slope

Figure 11.23   Formwork layout of residential tower

the overall building area was extremely large, the structure was subdivided into six main phases with further sub-sectioning in each phase. The main phases were constructed in a progressive and staggered manner both for the basement and the superstructure, each with a lapse of about one to two months. A large amount of construction jointing was provided during the construction, making the casting, coordination and quality assurance process fairly difficult (Figure 11.21).

4.2   Belcher's Gardens Redevelopment — using traditional manual-type timber form

The development is located on a sloping 24,000 sq m site. A 10-level podium structure, housing a carpark, shopping mall, and recreational facilities was built on top of the sloped site in order to provide a terrace to seat the residential towers. The upper and lower levels of the slope have a difference of about 65 m. Six 48-storey residential towers are located on top of the podium, providing about 2,200 residential units of 80 sq m each. To construct the podium on the formed slope, complicated falsework had to be erected, creating many elevated work positions that significantly retarded the progress of work. The huge size of the podium and the complicated site topography demanded very complicated phasing and sectioning arrangements (Figure 11.22).

As for the superstructure, due to irregularity in the layout, the incorporation of a lot of architectural features in the external envelope, and the use of a large amount of short-span slabs and shear walls, manually-operated timber panel forms were again adopted (Figure 11.23). As most structural forms for high-rise residential buildings have similar designs, the Belcher's Gardens redevelopment can be regarded as typical of these works in Hong Kong.

Figure 11.24   *Close-up of podium structure before construction of transfer plate for residential blocks*

Figure 11.25   *Falsework and formwork arrangement detail for residential blocks on a 2.5 m – deep transfer plate*

Other features related to the use of formwork:

a) The span of the podium averaged 12 m. Circular and squared section columns up to 3 m x 3 m in section were used (Figure 11.24). For the floor system, a flat slab design was employed for the lower floors to increase the headroom. A beam and slab system was used for the upper floors.

b) Inclined bracing beams were used in the upper podium structure to stiffen the residential tower on top.

c) A 3 m – deep transfer plate, tensioned and cast in 2 layers, was placed on top of the podium columns as support for the residential tower (Figure 11.25).

4.3   Lee Gardens Redevelopment — climb form for core, composite slab and structural steel outer frame

This is a 50-storey office building in the form of a composite structure with an inner core constructed of reinforced concrete and an outer frame of structural steel. The RC core was constructed using a VSL climb form, which is a self-lifting formwork system that uses hydraulic jacks to operate (Figure 11.26). The panel shutters used for the walls were operated on track rails to allow opening and shutting actions during the erecting and striking processes. The shutters and rail tracks, together with the scaffold

Figure 11.26   *Climb form for the construction of the core wall in the Lee Garden redevelopment project*

Figure 11.27   Overview of site layout arrangement during basement construction stage

Figure 11.28   Commencement of the core wall on its formation level

systems, were hung onto a steel gantry frame which further articulated the jacks to lift the entire system.

Other features related to the use of formwork:

a)   The 4-level basement was constructed in complicated phases using the top-down method with the old 2-level basement structure of the previous Lee Gardens Hotel carefully replaced during the process (Figure 11.27). Traditional manually-operated timber forms were used for basement construction.

b)   The basement portion of the core wall of the 50-storey office tower was constructed using traditional timber forms. The VSL climb form was erected after the completion of the ground floor slab and was used to cast the upper portion of the core wall up to the 50th floor (Figure 11.28).

c)   The construction cycle of a typical floor averaged 4.5 days per floor. Expected delays occurred at several locations, including on floors with outrigger provisions on which a very complicated anchor steel frame was required for insertion into the core wall to connect the outriggers, as well as on floors where the size of the core wall was progressively reduced.

4.4   Cheung Kong Centre — jump form system for the core; composite slab and concrete-filled steel tube for outer frame

Cheung Kong Centre is a 62-storey office building featuring a composite structure similar to the Lee Gardens Redevelopment but on a larger scale. Instead of using the climb form, this project employed a jump form system patented under the product name Cantilever (Figures 11.29, 11.30).

Figure 11.29   Jump form for construction of the core wall in Cheung Kong Centre project

Figure 11.30   Detailed set up of the jump form system

Figure 11.31   An outrigger system in the superstructure of Cheung Kong Centre ties the external building frame stiffly onto the building core

Besides using the form to construct the core wall in an advanced phase with the necessary provisions of starter bars for connection to the composite floor slab, the other difficult part of the formwork process was allowing placement of three sets of anchor frames inside the core for connection to 550 tonnes of outrigger frames at three prescribed levels (Figure 11.31).

Other features related to the use of formwork:

a)   The entire core wall structure for the 62-storey office tower was constructed using a jump form system which was erected for the casting of the wall starting from the lowest basement level. This was made possible by the provision of a 37 m diameter shaft for this purpose (the shaft was also used to facilitate the construction of the basement).

b)   The construction cycle for the typical floor averaged 3 days per floor. Expected delays occurred at several locations, similar to the problems encountered on the Lee Garden Redevelopment project.

4.5   The Gateway II — climb form for core and table form for slab

There are three towers in this project, each rising 38 storeys of housing offices and serviced apartments. The tower structures comprise a central core with 12 m – span RC columns placed to form the outer envelope and a post-tensioned flat slab system for floors and horizontal restraint. The core walls were constructed using the VSL climb form. Columns were built with gang forms and the slabs were cast using an aluminium-strutted flying form system (Figures 11.32, 11.33). The slabs were cast in two separate sections in a staggered manner with a lapse of two to three storeys (Figure 11.34). This arrangement has the flexibility of gaining one more work front so that the floor area can be split into smaller, easily-handled portions with better access for

Figure 11.32   Table form system used in Gateway II project

Figure 11.33   Work arrangement of table form and climb form

tensioning works. The drawback is that the integrity of the structure is broken and a number of construction joints were required in the structure. Besides, the number of sets of formwork to be provided, and the strength development period of the concrete before tensioning can be applied, had to be properly balanced in order to achieve the most efficient schedule.

4.6   Harbourfront Landmark — steel panel form for shear walls and table form for slab

This 72-storey residential development contains three residential towers on a 6,500 sq m site (Figure 11.35). The building structure consists of a series of shear walls forming the apartment units, centred with a core structure. A full steel form system was used for all the walls (Figure 11.36). For the slabs, an aluminium-strutted flying form system was used for the majority of areas except for a small portion at the rear housing kitchens and minor lobbies where traditional timber/plywood formwork was used. The central cores were constructed in the form of a vertical shaft. The inner structures such as the slabs for the lift lobbies, the lift walls and landings for the stairs were cast in situ at a later phase (Figure 11.37). The stairs were prefabricated and erected at pre-arranged positions inside the core.

Figure 11.34   External view showing the phasing arrangement of formwork

Figure 11.35   External view of Harbourfront Landmark during its construction

Figure 11.36   Steel panel form with the attach-on scaffold for construction of external wall

Figure 11.37   Layout arrangement of formwork for wall system around staircase core (Note also the placing in of a precast stair flight into the core wall shaft)

Figure 11.38   Layout arrangement of formwork for a typical floor in the Park Avenue project

Though the construction concept looks quite typical and similar methods have been widely used in public housing development in Hong Kong, this project is in fact a pioneer among private development. This is especially notable because of the complicated external shape of the building. The project was not run under a design and built contract; yet certain design alternatives were introduced. Within the relatively conservative culture of Hong Kong's construction industry, this may serve as a prototype in the application of more innovative technologies in private sector building construction.

4.7   Park Avenue (Phase II) — aluminium form

The project comprised four 46-storey residential towers (Figure 11.38). A full aluminium MIVAN formwork system was used in the construction (Figure 11.39). Typical of Hong Kong residential building designs, the structure includes a large number of shear walls forming the external walls, staircases and lift walls, as well as the majority of inner walls between apartments or other functional units. Beams serve as tie elements while external architectural features such as planter boxes and air-conditioner hoods are also incorporated into the design.

The first five typical floors took an average of 15 days each to complete. It then took about 8.5 days per cycle to complete each remaining storey. This is much too slow when compared to similar projects with floor cycles of five to six days. The problems seemed to come from the large amount of walls to be formed, a complicated layout and other architectural features, as well as inconsistency in the sizes of major elements. As a result, the labour input was unexpectedly high and caused significant delay to the works.

Other features related to the use of formwork:

Figure 11.39   Formwork arrangement for typical wall and floor (interior detail)

Figure 11.40   Falsework provided before erection of formwork for transfer plate

a) Supporting column clusters were first constructed from the ground to support the transfer plate structure, which was located at about 20 m from the existing ground level (Figure 11.40).

b) The support steel frame was erected on top of the columns cluster as falsework for the construction and casting of the 2.8 m – thick transfer plates (Figure 11.41).

c) The podium structure below the transfer plates was constructed afterwards from the ground level at the same time as the superstructure of the residential towers (Figure 11.42).

Figure 11.41   Completed transfer plate structure

5   Application of Innovative Formwork Systems in Hong Kong

Are the formwork systems employed in Hong Kong innovative? What is the future of more innovative formwork systems in the local construction industry?

According to some studies in this area, innovative construction technologies have the following in common:

Figure 11.42   Superstructure of residential blocks and podium structure below being constructed at the same time

a) They engage fewer resources to achieve the same output.
b) They yield better results when compared to traditional methods.
c) They have the ability to cater for other associated works during the construction process in a more coordinated manner.
d) They have better adaptability to cope with variances and changes in design.
e) They can complete the task neatly and faster.
f) They have safer work processes.
g) They are more environmentally friendly.

However, several basic conditions must be met before innovative technologies can be applied. Some of these conditions are given below.

External conditions (at community level)

a) Readiness and flexibility of procurement formats for construction projects, in particular large and complex ones;
b) Readiness of the related professions, which include developers, architects, engineers, contractors, sub-contractors and suppliers, to accept changes;
c) Expectation of society in terms of quality, cost effectiveness, wiser and greener products and services, etc;
d) Development of a mature market for the economical supply of new products and services.

Internal conditions (at institutional or corporate level)

a) Readiness for cultural reform in the search for excellence and quality performance, not just the adoption of window-dressing type exercises;
b) Development of better manpower with the required vision, experience, qualification and competence for technological and managerial advancement;
c) Development of the required support in terms of information technology, computer networking, equipment and other logistic support;
d) The managerial structure's ability to cope with new technology, market and other business environments.

The current situation in Hong Kong

The support of a very active economy over the past one or two decades has led to significant developments in the construction industry in terms of experience and the mastering of the required managerial, construction and engineering skills for handling very large and complex projects. At the same time, the motivating factors highlighted above have created an eagerness and readiness within the industry to advance. From the building construction point of view, the use of better formwork systems is no doubt a very direct way of introducing innovative methods to the construction of buildings. Below are some motivating factors:

a) Formwork labour cost is so immense that any innovative system resulting in a labour cost reduction is highly tempting.

b) Fulfilment of fast track construction schedule provides fewer choices, one of which is to adopt more innovative formwork systems.

c) Traditional systems can hardly satisfy the tight quality standard that is required nowadays.

d) Similarly, traditional systems can hardly satisfy current safety and environmental standards.

e) The accumulation of experienced operators makes the application of more sophisticated formwork systems more reliable and economical.

f) Many developers view the application of innovative technologies in the construction process as a positive image-building exercise.

Advanced formwork systems are only part of the advanced technology equation. Quite a number of recent projects have already integrated the use of advanced formwork systems and prefabrication techniques with success. Examples can be found in the Harbourfront Landmark project in Hung Hom as previously highlighted. Other examples include the construction of most public housing buildings such as Harmony and Concord blocks (Figure 11.43), staff quarters in Tai Kok Tsui and Lai King for the Hong Kong Police Force (Figure 11.44), some private commercial and residential developments, e.g. Swire House Redevelopment located in Central (Figures

Figure 11.43   Formwork arrangement incorporating precast wall elements in typical public housing projects

Figure 11.44   Government quarters in Tai Kok Tsui

Figure 11.45   Swire House Redevelopment project — overall set-up arrangement of wall and floor formwork

Figure 11.46   Climb form system to construct core wall

Figure 11.47   Table form for floor system butting onto core wall

Figure 11.48
An innovative jump form system in a recent residential development project

Figure 11.49   Details of the jump form as seen on the deck level

Figure 11.50   Kowloon Motor Bus Depot in Lai Chi Kok using a 70% precast structural system in construction

Figure 11.51   Details of precast elements forming the structural frame of the depot building

11.45, 11.46, 11.47), residential development in Stubbs Road (Figures 11.48, 11.49), or the depot building for Kowloon Motor Bus Ltd. in Lai Chi Kok (Figures 11.50, 11.51). These are projects that employed a large proportion of innovative elements in their construction.

The following are, in the author's view, the potential and limiting factors of innovative technologies in the built environment of Hong Kong.

Potential

a) The public's expectations (government, developers, building users) are rising all the time.

b) More stringent regulations have been enforced to control the performance of the construction industry.

c) Accidents are costly, especially where human casualties are involved.

d) The development or importing of more advanced technologies have become more common and affordable.

e) Some other work systems and supporting logistics are becoming more mature.

f) The industry is gradually accepting the production of higher performance buildings involving a more expensive resource input.

Limiting factors

a) Research and development at most contracting firms or other supporting units is insufficient.

b) Working space on construction sites (both on site or other work areas off-site) is inadequate.

c) Training opportunity (including on-the-job training) is still limited for both professionals and workers.

d) There is no guarantee of a stable market environment for the development and continual application of innovative technologies in construction (learned skill and experience will be lost eventually).

e) The extensive use of cross wall design especially in most residential buildings and small-scaled projects makes the use of more innovative formwork system less feasible.

f) The exceptionally large scale and complex nature of projects in terms of site condition as well as structural and building design confines the application of more advanced and sophisticated formwork systems.

The economic downturn in Hong Kong also means that neither developers nor contractors have the capital to invest in more innovative technologies. Similarly, the shrinking of the property market has created extremely keen competition that has discouraged the application of initially more costly innovative technologies in construction. This also applies to the supplier markets through which innovative products are often introduced.

Without the guarantee of a stable market environment for the development and continual application of innovative technologies, learned skill and experience cannot be easily accumulated. This, to a certain extent, can also explain why the use of structural steel in construction as a more innovative method of construction has not been adopted by local practitioners even though it is popular in Japan and other developed countries.

It is a pity that Hong Kong's construction industry may miss the chance to upgrade itself in the application of more advanced and innovative technologies due to the global economic downturn. When the recovery comes, it will inevitably take several years to build up the momentum for innovation. In the meantime, the industry or individual corporations may consider the following measures:

a) Explore ways to streamline and re-engineer the work structure at both the industrial and corporate levels.

b) Invest steadily in human resources development to train up more competent and high quality staff with the required aptitude and readiness to work in a new environment.

c) Invest steadily in the research and development of technologies that are particularly suitable for the built environment of Hong Kong. The "Integer" Project sponsored by the Hong Kong Housing Authority, China Light & Power, and Gammon and Swire Properties is a very good recent example.

d) Strengthen the linkages among government departments, developers, consultants, and contractor firms in the promotion, development, cooperation, and implementation of more innovative projects.

e) The government or other public institutions may consider providing funding to support research and development for the exploration, recommendation or setting up of guidelines and standards in the application of newer technologies and work systems in construction.

6   Conclusion

The selected cases in Section 4 illustrate the use of major formwork systems in various common construction scenarios in Hong Kong. The intention of this chapter is not to provide a detailed comparison or explain the technical features of individual formwork systems in depth. Instead, it aims to show the conditions and constraints governing the use of suitable formwork systems under typical local circumstances. There is no simple, readily-available solution for the use of formwork for complex buildings, especially when many Hong Kong projects are fasttracked.

The local industry has for a long time lacked the motivation to introduce highly innovative building methods due to a lot of understandable reasons. These include short-sightedness on the part of both developers and contractors regarding investment in research and development, an extremely competitive environment based on lowest-bid regimes, very high labour costs, and the relatively conservative culture within the related professions and industries.

The use of formwork in construction occupies a critical place in the technological improvement process. Yet, in this regard, the pace of change in Hong Kong has been rather slow. The economic downturn and restructuring as well as rising environmental concerns have provided the motivation to seek more efficient and higher quality construction systems. These adjustments can be as simple as improved training and attitude of the work team from the management down to individual labourers, their sense of loyalty and belonging, housekeeping issues on site, or safety and quality consciousness. These issues share equal importance in the introduction of advanced technologies as a whole. However, the slow adoption of these principles by Hong Kong's construction industry means that it has quite a long way to go before any genuine, long-term breakthrough is realized.

References

1. Hurd, M. K. 1995. *Formwork for Concrete 6th edition*. American Concrete Institute.

2. Irwin, A. W. and Sibbald, W. I. 1983. *Falsework — Handbook of Design and Practice*. London: Granada.

3. Steele, J. 2001. "Planning and Managing Innovation and Diffusion in Construction." *Innovation in Architecture, Engg & Construction*. Dept of Innovative Construction Engg, Loughborough Univesity.

4. Wilshere, C. J. 1992. *Formwork*. Thomas Telford.

5. Wong, Raymond W. M. 1998. "Battle within the ground — Experience learned from the Lee Garden, Hotel Redevelopment Project." Paper presented at the 5th International Conference on Tall Buildings, July 1998.

6. _____. 1998. *15 Most Outstanding Projects in Hong Kong*, China Trend Bldg Press.

7. _____. 1999. *Construction of Residential Buildings — Developments and Trends in Methods and Technology*, Hong Kong Housing Development, Book 2, China Trend Building Press.

8. _____. 1999b. "Common Formwork Systems," *Construction & Contract News 1 (1999)*. China Trend Building Press.

9. _____. 2000. "Prefabricated Construction," *Construction & Contract News 3 (2000)*. China Trend Building Press.

10. _____. 2000b. The Construction of a Semi-Buried Building — A Super- Sized Shopping Mall: The Festival Walk, International Conference Megacities, Dept of Architecture, University of HK.

11. _____. Feb 2001. The construction of the 62-storey Cheung Kong Centre. Newsletter. The Chartered Technical Review Paper. Institute of Building (Hong Kong).

Chapter 12

A Review of Common Technology for the Construction of High-rise and Complex Buildings in Hong Kong

Raymond W. M. WONG

Given the hilly terrain and extremely congested environment, Hong Kong may be regarded as one of the most difficult places in the world for the construction professional. Constructing a 40-storey building with a 4-level basement close to a busy mass transit railway tunnel in the centre of the city, or forming a vertical cut along a steep slope in order to construct a semi-basement type podium with a series of building blocks above it are typical examples of the extremely complex site conditions. This chapter highlights the difficulties faced by Hong Kong's construction industry and the techniques they employ to construct under such an environment. Emphasis is placed on construction technology employed by local practices.

PART III

CHAPTER

12

A Review of Common Technology for the Construction of High-rise and Complex Buildings in Hong Kong

1 Introduction

As an international city, Hong Kong is famous for its crowded urban environment and hilly backdrop. Within the 1,050 sq km territory, there are about 240 outlying islands occupying one-third of its total area. The remaining, less than 700 sq km of land, has to accommodate a population of 6.8 million people. This figure includes a series of mountain ranges that stretch all over Hong Kong.

In order to accommodate the huge population and provide the infrastructure and community facilities needed for an acceptable standard of living, commercial operations, as well as future development, many large, high-rise buildings are built at critical locations which are unsuitable for development by international standards. The following are some of the situations that the construction industry faces:

a) constructing a very tall building, often with a deep basement, in an extremely congested urban environment (Figure 12.1);

b) building in close proximity to very steep, or sometimes quite unstable, slopes (Figure 12.2);

Figure 12.1   Typical congested site with complicated layout environment (Lee Garden Redevelopment project in Causeway Bay)

Figure 12.2 Working on sloped site within urban environment

Figure 12.3 Constructing a large underground station in the new MTR's Tseung Kwan O Extension Line

c) building in close proximity, or sometimes even within, very sensitive and congested underground facilities like the Mass Transit Railway subways, surcharged areas of building foundations, or layers of large-sized drains, gas and water pipes, culverts, etc. (Figure 12.3);

d) building in close proximity to very large and complex traffic interchanges or busy transport facilities; and

e) building in newly reclaimed land or dump-filled areas (Figure 12.4).

Figure 12.4 Working very close to a seawall

Years of working within so many environmental constraints have enabled construction professionals in Hong Kong to adapt their practice to suit the local context, utilizing the technology and resources available locally.

2 Common Structural Forms for High-rise Buildings in Hong Kong

Structural forms adopted for high-rise buildings in Hong Kong are in fact quite limited owing to the following reasons:

a) land, planning and design regulations;
b) scale of development;
c) traditional design and construction practices;
d) efficient use of local labour and contracting expertise;

Figure 12.5   *Typical shear wall structure as used in public housing*

Figure 12.6   *Typical core plus external frame structure as used in most tower-type office buildings*

Figure 12.7   *Typical core plus external steel frame (composite structure) as a structural form: International Finance Centre Phase II Project*

e)   marketing trends that reflect the demand of end-users and the maximization of profit by developers; and

f)   design trends and sales strategies.

Popular structural forms employed for high-rise buildings in Hong Kong can be summarized as follows:

a)   In-situ reinforced concrete (RC) frame: This is the most popular system in Hong Kong. While the usual spans for such buildings range from 4 m to 10 m, depending on a building's design and function, spans of more than 20 m have been used recently, the majority of which are tensioned in order to minimize the size of beams. Beams are traditionally used for horizontal stiffening in a framed structure. However, flat slab structures which are often post-tensioned are growing in popularity, especially in commercial buildings because they provide a clear ceiling void to accommodate services.

b)   Load bearing or shear walls that replace columns: The number of beams used is often reduced to avoid the need for complicated formwork for forming the junction between beams and walls. Usual spans range from 4 m to 8 m. Due to the limited layout arrangements of space confined by load bearing walls, buildings using this structural form are commonly limited to residential apartment buildings. Panel type or large-sized shutter forms for walls and table forms for slabs are often used because of the highly repetitive nature of the construction process. However, details such as junctions between walls and slabs and staircases still impose a certain complexity on the design and erection of formwork (Figure 12.5).

c) In-situ RC (i.e. reinforced concrete) core wall with RC external frame: This structural form is used mainly in commercial buildings. The core wall which accommodates the lift shafts, staircases, toilet facilities and other building services provisions is usually square or rectangular in shape to facilitate the forming process. Wall thickness ranges from 0.6 m to 2.0 m, depending on the height of the building and loading requirements. Where there is no complicated architectural feature and the construction is highly repetitive in nature, large-sized panel shutter forms, sometimes mechanically self-lifting, are used to form the core structure. The majority of external frames are formed using more traditional, manually operated panel-type formwork, constructed at the same time or in a separate phase from the core wall structure. Spans ranging from 10 m to 12 m are the most common (Figure 12.6).

d) In-situ RC core wall with structural steel external frame (composite structure): This is similar to the above form but with the external frame constructed in structural steel. Almost without exception, the core wall of this type of building is constructed using a self-lifting formwork system ahead of the slab, with the connection to the steel frame to follow. A three-day per floor cycle can be achieved with a floor area of up to 2,000 sq m. However, the incorporation of an anchor frame in the core wall, especially on floors where bracing members or the outrigger frames are located, often complicates the construction process and retards overall progress. Effective spans range from 12 m to 15 m (Figures 12.7, 12.8).

e) Mega-structures using pure structural steel frame: This kind of structure is not too common because it is not rigid enough to cope with strong wind load under typhoon conditions, which occur in Hong Kong during the summer period. To strengthen the structure, complicated stiffening members in the form of transfer trusses, sectional floor plates, outriggers or heavy-sectioned bracing members are often required, which makes the design, fabrication, handling, and erection more difficult. However, composite designs such as composite floor with reinforced concrete topping or concrete filled columns are often used to increase the rigidity of the building. Recent examples of this type of structures in Hong Kong are the 70-storey Bank of China Building and the 80-storey "The Centre", which were completed in 1990 and 1998 respectively (Figure 12.9).

Figure 12.8   Another core plus external steel frame structure — Cheung Kong Centre Project

Figure 12.9   A pure structural steel frame structure (mega-structure) — The Centre

Figure 12.10   *Precast beams and slabs forming the podium of a low-rise building structure*

Figure 12.11   *Driving H-steel pile using track-mounted driving machine with noise reduction driving hammer*

Figure 12.12   *Driving H-steel pile using traditional equipment (static-type driving rack)*

f)   Semi-fabricated structures with precast concrete components: It is not too practical for high-rise buildings to be constructed using totally precast components because their relatively flexible nature are not suitable for typhoon conditions in Hong Kong. However, as a way to minimize the intensive use of expensive labour, semi-fabricated structures which use a certain number of precast concrete components are growing in popularity. In this case, the main structural members such as the core walls, columns, and main beams are cast in situ, often with some kind of patented metal forms. Secondary members such as stairs, secondary beams, short span tie beams, slabs (or semi-slabs) and external façades are constructed using precast methods. Most precast elements are placed with built-in link bars and cast at the same time as the main elements, to improve the rigidity of the joints. Post-tensioning is sometimes employed to increase the overall performance of the structure (Figure 12.10).

3   Foundation Systems and Methods

In addition to their height and the corresponding load they impose, buildings in Hong Kong have to cope with severe typhoon conditions which impose very complicated loading effects on the foundations of buildings. Wind speed above 200 km/hr is not uncommon during the summer typhoon season. Construction is also affected by features unique to Hong Kong, such as hard rock that lies close to the surface, subsoil with large amounts of boulders of a volcanic nature, work sites that are very close to developed areas with sensitive and congested underground or above-ground structures, the need to work under extremely tight schedules and complicated phasing arrangements.

The following foundation methods have been in use for many years and proven to be quite effective under local conditions:

a) Steel H-pile: Standard universal sections are used as piles with the load taken up by end-bearing and skin friction. The installation and equipment requirements are relatively simple, but the noise and vibration generated have to be restricted, especially in urban areas. In case of boulders, pre-drilling can be used before the insertion of the pile. This method is economical and effective for buildings of up to 30 storeys or above (Figures 12.11, 12.12).

b) Precast concrete pile: Precast pile can be square or circular in section. Prestressed hollow-section circular piles 10 m or 12 m in length are becoming more common in Hong Kong due to convenience, reliability, and cost-effectiveness. However, the method has major drawbacks, such as occasional concrete failure during the driving process, smoothness of pile surface reducing skin friction, as well as noise and vibration (Figure 12.13).

Figure 12.13   Using precast concrete pile as foundation

c) Mini-pile or pipe-pile: By using compact drilling machines, steel pipes of 150 mm to 250 mm diameter are inserted into the ground and grouted as pile. Various loading requirements can be met by controlling the number of piles used or by adding the number of reinforcing bars in the pipe before grouting. Due to the small diameter, drilling can be done fairly easily and causes only limited disturbance to the neighbourhood. This kind of foundation is suitable for use in congested areas with restricted working space or headroom. These piles can also be tensioned and can provide very good resistance to overturning due to wind load (Figure 12.14).

Figure 12.14   Forming mini-pile in congested working space

d) Small to medium sized in-situ concrete pile: piles ranging in size from 300 mm to 900 mm in diameter require the use of drilling

Figure 12.15 *Forming medium-size in-situ bored pile using continual flight auger*

Figure 12.16 *Forming medium to large sized in-situ bored pile using auger and bucket*

Figure 12.17 *Forming medium-size in-situ bored pile using track-mounted drilling rig*

rigs of an appropriate capacity for the bore. The drilling process can be facilitated by the use of steel casings or bentonite drilling fluid. Due to the rapid development of a wide range of highly effective mechanical drilling equipment, this foundation method is becoming quite popular for the construction of medium to high-rise buildings in Hong Kong (Figures 12.15, 12.16, 12.17).

e) Large diameter concrete bore pile: The boring process can be done manually or mechanically. In general, piles ranging in size from 1 m to 3 m diameter can be formed by mechanical methods while piles of 3 m diameter and above are dug using manual methods. However, in 1998 the hand-dug method was banned due to the high accident rate, except where special approval is obtained and certain safety requirements are met. Mechanical boring methods include the grab-and-chisel or reverse circulation drilling methods, both of which require the use of a steel casing to stabilize the bore during excavation. Sometimes, super large-sized piles of up to 6 m to 8 m diameter can be constructed. In this case, a cofferdam formed by sheet piles, soldier piles or in-situ concrete piles is provided for soil retaining (Figures 12.18, 12.19).

4 Basement and Substructure

In addition to the extra building area obtained by the provision of a deep basement, substructures can provide very good buoyancy to relieve the dead load of superstructures and counter-balance uplift due to wind load.

Figure 12.18   Forming large diameter bored pile using reverse circulation drilling method

*Figure 12.19
Forming large diameter
bored pile using grab
and chisel*

A cut-off wall is a major element in the excavation and construction of basement. There are many types of cut-off walls, the choice of which depends on a variety of factors, such as the scale and depth of the basement, the period of work, neighbourhood environment, mechanical equipment available, basement construction methods and cost planning requirements. Below are some common cut-off walling systems in use in Hong Kong.

a)   Steel sheet pile wall: This is most efficient for excavation to a depth of 8m to 10m. However, complicated horizontal support in the form of struts or bracing frames may be required which restrict further excavation. It is not suitable for use in areas with a large amount of scattered boulders (Figure 12.20).

Figure 12.20   Cut-off arrangement using typical sheet pile wall

b)   Soldier pile: This is similar to a sheet pile wall but H-piles are inserted into the ground at intervals with lagging structures to seal up soil surfaces between the piles. This is particularly suitable in areas with boulders because pre-drilling can be carried out in a fairly convenient manner. The H-piles can also be inserted into a concrete bored pile to produce a two-stage retaining design (Figure 12.21).

c)   In-situ concrete pile wall: Concrete bored piles ranging from 0.9 m to 1.5 m in diameter

Figure 12.21   Detail of soldier pile wall as an excavation cut-off using H-pile and steel sheet-pile as lagging

Figure 12.22   *Cut-off arrangement using caisson wall*

Figure 12.23   *Cut-off arrangement using pipe-pile wall*

are often used. The piles can be arranged in secant, continuously or at spaced intervals. Effective retaining depth can be up to 12 m or above and is suitable for use in more sensitive ground conditions because it is a vibration-free drilling operation (Figure 12.22).

d) Pipe-pile wall: This is similar to an in-situ concrete pile wall but smaller pipe piles are used. This system is most effective for use in very delicate environments where disturbance has to be kept to a minimum, or for small sites where large machines cannot be conveniently operated (Figure 12.23).

e) Diaphragm wall: Panels of trench walls are first formed by grab and chisel or by semi-automatic trench cutting machines. Wall thickness usually ranges from 0.9 m to 1.2 m. The forming of the diaphragm wall requires a significant amount of plant and may not be economical for small projects. However, its cost effectiveness increases with large sites which require a deeper retaining wall because it eliminates the need for complicated shoring supports for the excavation. A diaphragm wall can also be used as a permanent wall, to save fixing of formwork within the confined space of an excavated pit (Figures 12.24, 12.25).

Figure 12.24
Cut-off arrangement using diaphragm wall and laterally restrained by ground anchors

Figure 12.25
Forming the trench for diaphragm panel using clamp shell

In addition to the use of an appropriate cut-off wall to facilitate excavation, grouting is often used for subsoil strengthening and to improve the ground's water-tightness. As the excavation proceeds, horizontal support using strut frames or other shoring and bracing systems need to be installed to counteract the lateral pressure on the new cut. Sometimes, where the situation permits, ground anchors instead of a strut system can be used as lateral support, to gain more working space inside the work pit. Finally, there should be appropriate dewatering provisions to suit the specific geological or neighbouring environment, to keep the pit safe and free from groundwater intrusion.

Figure 12.26   Lateral support system using sheet-pile and strut frame for basement constructed under bottom-up arrangement

Basements can be constructed using the traditional bottom-up method or the top-down method. The procedure of the two methods can be summarized as follows:

Bottom-up method

a) Construct the cut-off wall.
b) Start excavation within the basement parameter.
c) Erect lateral support layer by layer as excavation proceeds until the required depth is reached.
d) When the formation level has been reached, construct the foundation rafts, pile caps or ground beams.
e) Construct the basement slab and other internal structure starting from the lowest basement level. The works are usually done in carefully scheduled sections to avoid the disturbance of the strut members.
f) Repeat the basement works until it reaches ground level.
g) Release and dismantle the strut members at suitable stages as the basement structure is completed.

Figure 12.27   Forming the pile caps and ground beams in the formation level of the basement

Figures 12.26–12.28 show the construction of a basement using a typical bottom-up approach for the Lee Theatre Redevelopment project.

Figure 12.28   Forming the basement slab under typical bottom-up approach

Figure 12.29 The core wall and the column structures
              being erected before the basement are
              constructed in a top-down manner

Figure 12.30 Carrying out the excavation
              within the top-down basement

Figure 12.31 Forming the floor slab inside
              the basement interior

Figure 12.32 Making use of the "double bit" method
              to cast an intermediate basement slab

Figure 12.33
Erection of the
reinforcement bars
before the encasing
of the basement
columns using
concrete

Top-down method

a) Construct the appropriate cut-off walling
    system (usually a diaphragm wall is used for
    this method as it is more effective in
    handling large-scale projects).

b) Erect temporary columns, usually in the
    same position as the permanent columns, in
    the form of steel stanchions, to support the
    basement structure to be constructed from
    the top level downward.

c) Construct the first basement floor slab,
    usually starting from the ground level, which
    will also be used as the horizontal support to
    the cut-off wall as excavation proceeds.

d) Excavate downward and construct the second level basement slab in the same manner. Erect intermediate temporary shoring support where required.

e) Repeat the excavation and basement slab construction until the required depth is reached.

f) Construct the foundation caps, rafts, ground beams, or sub-soil drains as required.

g) Construct other internal structure where required.

h) Encase the temporary columns in concrete to transform them into permanent columns.

Figures 12.29–12.33 show the construction of basement using a typical top-down approach for the Cheung Kong Centre project

The bottom-up method has the benefit of requiring fewer plant facilities to operate and is therefore more suitable for use in smaller sites where the cut is relatively shallow. However, its disadvantages appear where a site is bigger, the basement is deeper, or the project is under a tighter schedule. Under this situation, very complicated horizontal support has to be erected, thus increasing the working time and cost of the project while limiting the working space within the basement pit. Unlike the top-down method where construction of the superstructure and basement can proceed at the same time making use of the first basement slab as a separating plate, superstructure construction can only commence upon completion of the basement structure when the bottom-up method is used. Hence, basement projects of a large scale, almost without exception, are all constructed using the top-down method in recent years.

No matter which method is used to construct a basement, some very fundamental considerations must be given to certain critical factors in the planning and execution of a successful basement project. These include the detailed arrangement of the erection or later dismantling of the temporary support; sub-division of the basement structure to workable sections and the provision of construction joints; coordination between phased sections to cope with the access of labour, plant, and materials; spoil removal arrangements (Figures 12.34, 12.35, 12.36); and matching basement work with progress on the superstructure.

Figure 12.34 Spoil removal provision in basement construction using muck-out and grip

Figure 12.35 Spoil removal provision in basement construction using hoist and bucket system

Figure 12.36   Entrance provision into a basement constructed using top-down arrangement

5 Construction of Superstructure

Superstructure construction involves three major activities, namely:

a) the provision and erection of a suitable formwork system;
b) steel fixing; and
c) the placing of concrete.

In addition, an efficient site layout with the required plant facilities properly set up is a "must" if a project is to be successfully completed on time and within budget.

There are a wide variety of formwork systems available for high-rise construction. In Hong Kong, however, the traditional, manually operated timber form is still the most frequently used system due to its flexibility in handling difficult shapes and non-standard layout requirements without as much plant or logistical support as other formwork systems. Although it has obvious drawbacks, such as its labour intensive nature, slower speed of work, environmental unfriendliness, or unsatisfactory concrete quality, it is still used for many small-scale projects because of its simplicity.

For larger projects with tighter schedules and higher performance requirements, the following formwork systems are often employed:

a) large-sized panel shutters, often in mild steel or coated timber/metal combination (Figure 12.37);
b) manually operable aluminium panel forms for both walls and slabs (Figure 12.38);
c) table forms and flying forms for slab construction (Figure 12.39);

Figure 12.37   Large-sized panel forms to construct a core wall

Figure 12.38   Aluminium form system capable of casting the wall and floor in a continual process

Figure 12.39   A modified table form system for casting the floor slab of an office tower

Figure 12.40   A typical climb form system used for casting the core wall of an office building

d) self-lifting formwork systems such as slipforms, jump-forms and climb-forms, often of a special patented design (Figure 12.40); and

e) other patented formwork systems such as the SGB, RMD, or PERI systems.

Figure 12.41   Steel reinforcing bars being fixed in position for a complicated beam section

The practice of fabricating and fixing of steel reinforcement in Hong Kong is still very traditional (Figure 12.41). The current and most widely used method is to cut and bend the steel bars on site, transport the bars to the work spots by tower crane, and then fix the bars into position. The use of prefabricated components from the workshop for erection on site is uncommon. This can be explained by non-standardized design, use of slim structural elements and congested reinforcement design aimed at minimizing the size of members due to insufficient working space for placing in prefabricated bars.

Concrete with strength ranging from 20 to 35 N/mm^2 (Grade 20 to 35) is most commonly used due to easy performance control in a place where supervision standards and the quality of labour are not consistent. However, in order to produce higher and slimmer structures as well as to reduce the dead weight of a building, Grade 40 or even Grade 60 concrete is occasionally used. Typical examples are core walls, transfer structures and large-span or tensioned structures. Most of the concrete used is supplied by specialist suppliers and delivered to site by concrete mixer trucks.

Concrete is placed for high-rise buildings using mechanical equipment such as booster pumps and placing booms or skips, hoppers or

*Figure 12.42   Typical concreting arrangement for
              an extremely small and congested site*

*Figure 12.43   Typical set-up arrangement for placing con-
              crete using manual labour and wheel barrow*

*Figure 12.44   Placing concrete using tower crane
              and skip bucket*

Figure 12.45   Placing concrete using concrete pump

*Figure 12.46   Placing concrete using a hydraulic-operated
              booster arm*

buckets lifted by tower cranes, hoist racks or similar devices. The placing process may still be quite labour intensive because of the horizontal movement of concrete and the need to achieve satisfactory placing and compaction within a congested working environment (Figures 12.42–12.46).

In terms of speed of work, fast-tracking is very often mandatory under most contract requirements. In general, a floor cycle of five to six days is required for a typical 800–1,000 sq m building. For larger buildings with more complicated layouts, progress may take ten days or more for a typical

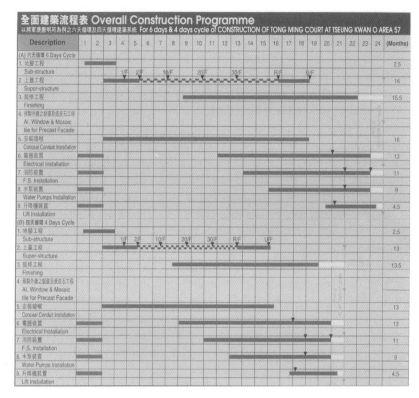

Figure 12.47   A typical bar-chart program showing the planning arrangement of a construction project

floor. Planning options include the sub-division of the floor area into two convenient phases, casting the walls first with the floor/beam to follow, or even working with a staggered floor arrangement, to obtain the best scheduling result and efficient use of resources given the layout and other design constraints (Figure 12.47).

6   Conclusion

The construction of high-rise buildings is a very broad subject. Within these pages, therefore, it is only possible to highlight a few of the local features. In addition to the application of technology in construction, the overall structure and common practices in the local industry are, in fact, the most important determinants of construction outcomes.

In recent years, the overall performance of the construction industry has improved due to a number of positive developments. One such development is the change in the property market from a sellers' market to a buyers' market following the economic downturn in 1997. It has led to the construction of higher quality buildings aimed at satisfying the genuine needs of the public. Some forward-looking developers are beginning to invest more in investigating into the possibility of and options for employing advanced technology that will make their buildings more reliable and perform better. More innovative and environmentally friendly design and construction processes have become a focus of development. Developers now demand, through the appropriate budgetary and contractual means, that planners, designers, and main contractors deliver projects of the desired quality. The construction industry has realized that it can no longer stick to the old ways.

At the same time, thanks to a change in attitude on the part of the government, which has become more answerable and accountable to the general public, a more rational control and implementation mentality is being adopted. A tighter control mechanism for building development, such as the incorporation of stringent submission requirements and monitoring of production control processes, has been introduced. At the same time, the government is becoming more flexible and open-minded towards newer and more innovative ideas in design and construction, thus indirectly encouraging the design and construction of better quality buildings.

As a result of such changes, smaller firms operating under traditional managerial systems with limited capital and resources can hardly survive, for these firms may not have the required competence to fulfil the current statutory requirements and performance standards as set down by the government for complex projects; nor can they compete with other more able counterparts for jobs. Construction firms of the new generation are becoming more eager to establish better work organization with stronger professional capability. They are more willing to invest resources in research and development and are more prepared to upgrade themselves by restructureing or setting up more efficient administration systems at various levels.

In terms of the application of construction techniques, the traditional, more labour-intensive methods can hardly meet the time, cost, quality and safety requirements nowadays. These construction methods will gradually fade out, except for very small-scale jobs where advanced construction techniques cannot be practically applied. The gap will be filled by industrial and rational approaches towards most development projects. This will lead to a wider use of system formwork and coordinated components and the use of more prefabrication, tensioning or automation techniques; on the design side, more standardized, modular and high-performance structures will be adopted together with long-span designs and high-strength materials. These advanced systems will require stronger and more competent support in terms of coordination and administration in order to achieve the expected result.

In the author's opinion, the gradual change in this direction is obvious and the process of modernization will lift the overall professional standard of the Hong Kong construction industry, probably within a decade or so.

References

1. Buildings Department. 1998. *Building Construction in Hong Kong 1998.* Hong Kong Special Administrative Region Government.

2. Fan, Andrew. 1998. "Construction of High-rise Concrete Buildings in Urban Areas." In *Building Construction in Hong Kong 1998.* Buildings Department, Hong Kong Special Administrative Region Government.

3. Gibbons, C., Ho, G. and MacArthur, J. 1999. "Developments in High Rise Construction in Hong Kong." *Proceedings of Symposium on Tall Building Design and Construction Technology.*

4. Grosvenor, R. 1998. "Comparison of Hong Kong & Overseas Construction Practice & Management." In *Building Construction in Hong Kong 1998.*

5. Wong, W. M. 1999. *Fifteen Most Outstanding Projects in Hong Kong.* Hong Kong: China Trend Building Press.

6. _____. 2000. *Construction of Residential Buildings — Developments and Trends in Methods and Technology.* Hong Kong: China Trend Building Press.

Chapter 13

Coordination between Building Services Installations and Construction Processes

Ir Lonnie CHEUNG

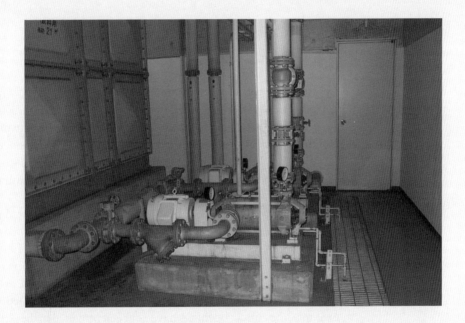

Improper coordination and poor programme management of building services installation may eventually affect a building's structure. Workmanship may be poor and, more significantly, there may be delays in the overall construction programme leading to additional financial claims and late handover. This chapter discusses the coordination requirements for air-conditioning, electrical, fire services and plumbing systems. Important aspects of building services installation procedures and their importance to a successful project are also discussed.

13 Coordination between Building Services Installations and Construction Processes

1 Introduction

Building services systems are frequently designed without due consideration for installation procedures. Once a contract has been awarded, the builder moves in and starts construction work on site. Building services installers usually move in at a fairly late stage. Building services installations are seldom planned ahead and the correct work sequence is only followed later in order to keep pace with the builder's progress.

Remedial works for building services, which are required more often than those for other trades in the building industry, are blamed for the late award of contracts and appointments as well as the insufficient time to organize and deploy engineering staff. In fact, excellent forward planning is just as important as excellent installation techniques, which alone cannot guarantee a building services system will be installed well if, for example, the correct openings have not been provided in the structure. The need for modification work when this happens inevitably lead to project delays. The overall construction cost will increase, and loss due to delayed occupation may also result, leading to liquidated damages and fines. Clearly, improper building services installation procedures can cause serious problems. An in-depth investigation of the origin of these problems is needed if a solution is to be found.

The design and installation of building services systems are two different stages of work used to be carried out one after the other. However, because building projects are so fast-paced nowadays, the two stages often commence simultaneously or with a minor time gap. The award of contracts to successful building services contractors is usually done at the last moment, when the contractor has already laid the concealed conduit and pipe work in order to avoid delays. In 1983, when the problem first surfaced, it only happened occasionally. In the early 1990s, however, the problem became worse as a result of control mechanisms imposed by project managers. Intended to solve certain problems, these measures gave rise to other problems.

A good understanding of building services installation requirements and proper installation management procedures is needed if the problems is to be identified at an early stage so that effective solutions can be devised to rectify the problems.

Building services contractors are normally engaged as sub-contractors by a main contractor, i.e. the builder who is responsible for completing the project. Building services cover four major trades, namely: heating, ventilation and air-conditioning system (HVAC); electrical services; plumbing and drainage services; and fire services. All four must be installed in an integrated manner within a building. The following is an analysis of the problems arising from poor coordination between their installation and structural works.

2 Heating Ventilation and Air-conditioning System (HVAC) — Coordination with the Construction Processes

Inadequate space given to plant areas for HVAC equipment is the first and most frequently encountered problem because this kind of equipment is usually bulky and heavy. They can produce considerable mechanical noise and vibration. Design engineers are usually faced with the difficulty of moving large equipment into plant rooms when the space provided is inadequate. Maintenance and removal are also difficult. Poor coordination at the construction stage may lead to completed structures, such as concrete slabs or beams, being opened up or cored, which in turn may affect structural integrity if the subsequent repair work does not actually restore the structure to the original design specifications.

To avoid this problem, design engineers should set aside a reasonable amount of space during the design stage. Careless or irresponsible planning could have a negative impact on the project and maintenance problems may occur later if inadequate space is provided. High land price has made architects less willing to assign more space for such equipment, but good engineers do not ask for more than what is adequate.

Large equipment demands large access routes and areas for moving and turning as well as installation and maintenance. These routes and areas may require large openings other than doorways or access hatches, but the need for them is not always identified by design engineers or inexperienced engineers, who should discuss access requirements with the builder. The construction sequences may have a major impact on the installation if no allowance for their movement and installation has been made in the building structure. Bitter experience from many cases indicates that the only solution is to break the slab, which may affect structural integrity if the subsequent repair does not actually restore the structure to the original design specifications.

Heavy equipment require strong hoists and tower cranes for transportation to the appropriate plant rooms in the intermediate floors and on the roof. Builders must make available lifting equipment capable of transporting such heavy equipment. At the same time, the HVAC contractor should plan the plant access route and avoid the dismantling of wall and structures, to avoid compromising structural integrity. Any damage to the structure may lead to water leakage through joints and dampness. There are cases showing evidence of these problems.

Installation procedures should be carefully designed during the planning stage of a project. The procurement and delivery of equipment at the right time is essential, and lifting equipment must be available for their transportation. By reducing noise, plinths and anti-vibration mounting devices are very important to the successful installation and operation of the equipment. The absence of such devices through negligence during the design stage may lead to expensive remedial work later. Engineers must also ensure that access routes are sufficiently wide, with enough space between equipment and walls or columns. Sufficient clearance between equipment and walls other than the normal openings is required; and equipment access route of sufficient width must be provided during the construction stage.

3 Electrical Services — Coordination with the Construction Processes

Electrical services are installed by connecting various devices to cables and bus ducts. Exposed cable trays as well as concealed conduits embedded in the concrete are used to lay cables and electrical wirings.

The concealed conduit system came into practice in the early 1970s. A review after years of operation shows that a small percentage is damaged or blocked when conduit joints are blocked by concrete loosened by the vibration employed for concreting. Totally concealed conduit systems with junction and terminal boxes make it very difficult for

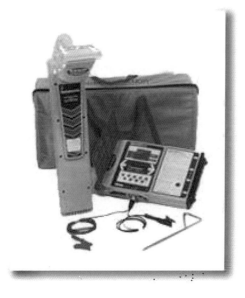

Figure 13.1   Locator (Courtesy of Radiodetection (China) Ltd.)

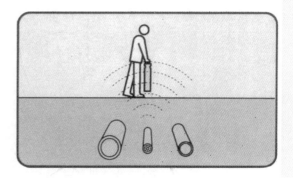

Figure 13.2   Use of locator to determine the conduit location in floor or wall (Courtesy of Radiodetection (China) Ltd.)

electrical workers to prevent concrete leakage into the boxes and conduits via the joints. Preventive measures and installation practice guides issued by local authorities and utility companies help reduce the occurrence of damage, but do not eliminate the problem observed.

Locating misaligned concealed conduits and blockages during construction and the subsequent rectification work may cause delay and increased cost due to the need to catch up with the building programme following rectification.

The frequent occurrence of this problem has led to the invention of a device named Locator (Figure 13.1) which is used for detecting metallic items inside walls or floors (Figure 13.2) and can identify steel members, water pipes, and conduits embedded inside concrete when openings need to be made in walls and slabs.

Another way to minimize the use of concealed conduits is the use of raised floor systems. Conduits and trunking can be flexibly located inside the raised floors, which will offer even more benefit if the air-conditioning system is also designed for raised floors.

Wireless systems can help reduce the use of cables and concealed conduits, but there is concern that infrared radiation and ultra-sonic wave penetration may lead to cancer. Until medical research indicates wireless communication is safe, a cautious approach towards its adoption is advised. Besides, wired communication systems are still superior to wireless systems in terms of speed and bandwidth.

Surface-mounted conduits covered by false ceiling systems is a practical alternative. Depending on the actual requirements of the developer, surface-mounted conduits not covered by false ceiling systems are rarely acceptable in office buildings of a high standard.

Electrical works must be tested to the satisfaction of local authorities before the relevant utilities completion certificates can be issued. Their completion may be affected by delays or defects in other building services. There are cases where a power supply connection has to be deferred because the installation of other services have been delayed due to technical faults in the electrical parts of HVAC services, fire services and/or plumbing and drainage services. These problems can adversely affect the timely completion of electrical systems.

Nowadays, developers like to employ coordinators to monitor work progress, but very often there is a lack of suitable building services coordinators, so both developers and contractors lose money in the end and find themselves engaged in arbitration and/or lawsuits.

4   Plumbing and Drainage Installations — Coordination with Construction Processes

Water marks and leakage affecting the floor below or the slab soffit are becoming a serious problem in both old and new buildings that require frequent repair work. The Buildings Department recommended the use of exposed potable water pipe at the end of 1999, but before then, small potable water pipes were typically embedded inside concrete, which could cause repair and maintenance problems at a later stage. Where remedial works are needed, concealed pipes could also cause delays in the work of a construction team.

Obviously, there are major differences between concealed pipework and exposed pipework installation, whether it is from the aesthetic or the maintenance perspective. As always, good-looking buildings are usually more difficult to repair or maintain. The building services installation is particularly for concealed water pipework.

In technical terms, there are major differences in the installation of concealed and exposed piping. The short notice usually given by builders for laying and fixing pipe work inside the formwork ready for concreting means workers often have to work overtime, even till midnight, to catch up with the work schedule. Poor illumination and lack of preparation at night could lead to work of an inferior quality, especially when tests have not been completed but the concreting schedule is adhered to because the vehicles bringing the ready-mixed concrete to site are waiting. The short notice would not be a problem if the sequence of work is better arranged. The problem has been partially solved by the newly released guidelines, but it can only be completely resolved through heightened awareness and standardization of work sequence.

A top management decision to pay more for the use of exposed installation could eliminate the problem. Of course, experienced coordinators could also foresee the problem and plan ahead accordingly. The guidelines from the Buildings Department are not enforced as law, so it will not prevent the problem from recurring. The government therefore may have to legislate sooner or later.

Repairing faults in concealed piping may rectify the problem but affect structural integrity and the surface finish of a property. Breaks in pipe work remain weak even after repair. Leakage may also recur, creating a leakage path and weakening the structure. This type of leakage is unpredictable and its cause has been identified as uncertain by those who are not familiar with it. It is a problem which engineers must pay attention to and further investigation and data gathering are recommended. Good record keeping can help trace the source of leaks and the leakage paths and thus the search for an effective solution.

Good construction programme planners and administrators are valuable assets to building services installers. They can devise and manage effective programmes and design effective work sequences with the builder, but many planners do not have the expertise to make allowance for contingency or precautionary work in the main programme. For example, typical foreseeable problems such as bad weather are unavoidable. Planners must allow some contingency in the main programme, for instance by using the weather data of the previous year as the basis for the development of a programme with a reasonable buffer, to prevent a late handover.

The way towards project completion can be littered with problems and remedial work. For example, the application of nails and wire meshes during the finishing stage may damage the concealed pipe work and delay the project. A newly developed locator (Figure 13.1) can be used to detect the location of drain and supply pipes during installation preparation (Figure 13.3), to determine the position of metallic items so that the installer can position them correctly (Figure 13.4).

In the middle of the construction phase and upon the commencement of building services installation, scaffolding and additional support are needed to hold the water supply pipes and drainage pipes in the appropriate positions. Cooperation between the different parties responsible for the different types of works is absolutely essential. The builder should avoid damaging pipe work when clearing or cleaning external works during the final stages of construction. The application of

Figure 13.3   Use of locator to locate water pipe or gas pipe
(Courtesy of Radiodetection (China) Ltd.)

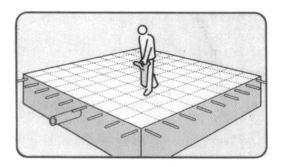

Figure 13.4   Use of locator to locate water pipe in floor slab
(Courtesy of Radiodetection (China) Ltd.)

acidic solutions, which harm such services installations, must be avoided. To eliminate the problem, the builder should be warned in writing at the beginning of the installation stage of the potential impact of such solutions. As a general rule, some of these procedures become a part of routine procedures. Mutually agreeable standard procedures and formalities are strongly recommended.

Damages to sanitary ware happen quite frequently. Although warnings may have been issued against damage during installation, there is no guarantee that damage will not result from the wet trades' work. Such damages may affect the associated services installations, and, again, repair works may have an impact on structural integrity. Delayed completion means late inspection by the Water Supplies Department. Under the current practice in private property developments, this may mean developers having to compensate property owners for late occupation.

5 Fire Services Installations — Coordination with the Construction Processes

Fire services installations face the same problems and demand the same attention as potable water systems, except that no fire services pipes are embedded inside concrete. Care must be taken to ensure that fire services pipes installed inside staircases will not affect their width as stipulated by the building codes.

Sprinkler pipes are distributed across the ceiling, secured by pipe sleeves precast in the beams and walls to allow the pipes to pass through. These sleeves must be positioned correctly if construction delay is to be avoided. The installation of sprinkler pipes also reduces the clearance of a room, so a suitable floor-to-floor height must be planned accordingly. Construction tolerance must also be taken into account.

The second major problem affecting fire services installation is coordination with other services occupying space in the ceiling. A higher ceiling requires two layers of sprinkler heads, one close to the structural ceiling and the other at the level of the false ceiling. This network of pipes prevents other services from running freely. The final location of the upper layer of sprinkler heads cannot be confirmed without coordination with other electrical and mechanical services installations. In fact, the ceiling installation sequence and the space required for carrying out the installation are vital to the successful installation, performance, and maintenance of building services systems, but they are usually not well attended to, thus giving rise to problems.

Coordination of fire protection work during construction is also important, particularly for steel structures, which must be coated with fire-resistant materials since they cannot resist fire on their own. Sometimes services installations are carried out rapidly without other parties being notified. Since structural fire protection has to be applied before services installations are carried out, building services which are already installed will have to be dismantled to enable structural fire protection work to proceed. All parties lose when this happens.

6 Conclusion

Better understanding of installation procedures and better preparation before and during construction are absolutely essential because they have a significant impact on the quality of the final works. Coordination means better arrangement of work schedules to avoid conflict during installation of building services systems. When a project team fails to pay sufficient attention to the coordination of the works, delays and/or financial loss may result. This may also give rise to future problems associated with the maintenance of the systems concerned. Figure 13.5 is a good example of proper provisions for a plant area. The opposite is illustrated in Figure 13.6.

In order to enhance installation works and extend plant life, adequate attention must be given to coordination of building services installation. Good coordination not only requires good planning, but also good coordinators. Good coordinators are those experienced engineers who possess sufficient understanding of various professions involved in the building process and are able to anticipate problems before they occur.

Figure 13.5   Example of good plant space allowance

Figure 13.6   Inadequate maintenance space

Epilogue

Bill LIM

This publication was compiled as Hong Kong entered its fifth year after its return to China in 1997. HKSAR (Hong Kong Special Administrative Region), as the territory is now known, has gone through a series of events that have brought about a certain degree of uncertainty. The new airport took some time to settle back to normality. The effect of the currency crisis which has affected most Asian countries is still being felt in Hong Kong. The collapse of the property market has brought about a class of property owners who are over-burdened by mortgage. Land sale is below expectation, both in price and in auction participation. Poor building practices such as short piles and landslides have given rise to severe criticism of government policy.

The same uncertainty is also observed as regional and international affairs unfold. The admission of China into the World Trade Organization (WTO) will lead to some fundamental changes in trade practices. As China is preparing for a large influx of foreign investment and a major increase in the export of goods, the traditional position of Hong Kong as an *entrepot* is being questioned. The proposed development of the city into a free trade zone and the eventual link-up with the Pearl River Delta will take some time to develop, given the lack of unanimity on 24-hour access across the border with Shenzhen. The tragic event of September 11 has drastically reduced the number of tourists from the USA.

Internally projected growth is almost zero in 2001. The Government is heading for an unprecedented budget deficit. Retrenchment of workers and premature closure of businesses are reported frequently. Requests for rental discount on shops and public housing are made on the ground of economic hardship. The Basic Law is being rigorously tested as immigrants argue for the right of abode in Hong Kong.

No doubt Hong Kong will rise from such uncertainties and recession. Hong Kong people are resilient and have overcome challenges more formidable than these before. The new airport is now one of the finest in the world. The currency crisis will eventually be over, as has already been witnessed in Singapore. The property market is possibly at its lowest ebb, and is likely to rise as Hong Kong is opening up to foreign investors and professionals with the required expertise and entrepreneurship. Shenzhen land is now seen as viable alternative to Hong Kong land, as the relationship between these two cities continue to strengthen. Professional ethics and acceptable practices being scrutinized by respective professional bodies will restore public confidence.

China's entry into the WTO is seen by many as a challenge instead of hindrance for Hong Kong. The administrative issues will eventually be resolved as both Hong Kong and Shenzhen appreciate the opportunities afforded by minimizing access restrictions. The greater economic zone of the Pearl River Delta, eventually linking Hong Kong and Guangzhou, will become a magnet of trade, manufacture, and commerce in South China. Signs of such development are already evident, for example the expansion of the insurance industry from Hong Kong into mainland China. Such economic activities will be closely followed by infrastructure including transport, utilities, and building development from factories to housing estates. The Hong Kong power grid is already linked with that from the mainland, and the water supply agreement between Hong Kong and mainland authorities including quality control is likely to be refined further.

Tourists from China are now much more common, and their number is expected to rise as prosperity continues under the prevalent policy of economic reform, which is likely to be further enhanced by the WTO admission. The Disney development is but one of the attractions for mainland Chinese, as Hong Kong offers a wide choice of consumer goods which are now affordable to many.

The recovery of Hong Kong's economy ultimately depends on her people's will to make it happen. No longer can the people of Hong Kong sit on their past laurels. Their world is much more competitive, especially against the backdrop of the impressive if not astonishing progress in China, of which Hong Kong is now a part.

To do so there must be a change in attitude, which can only be tempered by education. For education is not only to provide the people with the necessary skills, but also to convert the mindset from the passive to the active. Education, especially for the young, is essential to this conversion, as the unequipped are unable to grasp the opportunities presented to them.

In this regard, the Hong Kong government has provided the necessary leadership; for instance, re-training the less successful, intensifying language and numeric training, popularizing the use of computer — all these efforts aim to foster a proactive learning environment. In tertiary education, the government has taken the initiative to introduce associate degree (AD) programmes as an alternative academic path, leading to qualifications that will enable students to either develop a professional career or further their education upon completion of a two-year intensive programme. The government's initiative is a direct response to the growing demand for advanced technological education in Hong Kong. This is an innovative approach, and its success depends on the ingenuity of tertiary institutions to fully utilize the relatively short time available to equip students for the challenges and opportunities ahead.

Under normal circumstances, building activities follow economic recovery. Professional education must therefore begin prior to recovery to supply the necessary personnel that will eventually be required, as quality human resources need time to develop and mature.

In support of the government's initiative in higher education, the staff of the Division of Building Science and Technology are fully aware of their responsibility in supplying the building industry with qualified technologists. The problem-based teaching and learning approach is seen as a possible alternative to theoretical and idealized classroom learning through practical applicaion of knowledge in realistic situations that students are likely to encounter in their professional practice.

As mentioned in the Prologue, communication between the Division and the professional community is crucial for the educational endeavour to be successful. Professionals who contribute to the academic programmes as part-time lecturers bring with them real cases encountered in their practice. Accreditation exercises give quality assurance to the teaching and learning process. Conversely, academics participating in the various activities of professional associations gain insights into the future direction and progress of the profession.

As the first batch of 2002 AD graduates are now part of the workforce, their performance in the building industry will be the first test of the success of the AD programmes. Constant feedback from the profession and self-assessments by graduates will certainly contibute towards the future development of the programmes.

About the Contributors

Bill B. P. LIM
PhD, BArch (with Hons),
DipTCP SydUniv, FRAIA, RIBA,
(Former) Emeritus Professor,
Queensland University of
Technology, Australia

The late Professor Bill LIM was dean of architectural schools at National University of Singapore and Queensland University of Technology. He was an active organizer of international associations and conferences in architectural science. His main contributions to the field include building science, daylighting, high-rise building, environmental aesthetics, fire-proofing and traditional architecture.

Apple Lok Shun CHAN
MPhil CityU, CEng, MInstE,
MASHRAE

Apple CHAN is mainly involved in teaching environmental science related subjects. His research interests include energy audit and survey, energy conservation, building thermal analysis simulation and component-based HVAC plant simulation.

Ir Lonnie CHEUNG
MSc, AP(HK), MHKIE, CEng,
RPE(BS), MCIBSE, MIHEEM, MIP

Lonnie CHEUNG has worked in the building services industry for over 23 years, serving mainly in the consulting field. Lonnie has been involved in famous and high-rise building systems design and complex services coordination work in Hong Kong. He also provides expert advice to professional institutions and regularly presents his research findings at academic and professional conferences in Hong Kong and China.

Tin-tai CHOW
MSc(Eng) HK, MBA CUHK, PhD Strath.,
CEng, MIMechE, MIHEEM, MCIBSE,
FHKIE, MASHARE, RPE (BS,MCL)

Tin-tai CHOW is a Fellow Member of the Hong Kong Institution of Engineers. He has published widely on building science and technology subjects. His expertise is in building environmental technology and energy studies. He is particularly interested in the application of simulation techniques in solving practical engineering problems

Jackson KONG
BSc(Eng), MASc, MEng, PhD,
CEng, MHKIE(Civil), MIStructE,
RPE(Civil)

Jackson KONG has practiced in Hong Kong and Canada in various areas of civil engineering. He has been involved in the design of prestressed concrete bridges, including the recent award-winning West Rail Viaducts. His research interests include computational mechanics, bridge engineering and micro-tunneling. Currently he is involved in a research project on the analysis of cable-stayed bridges. He has published numerous papers on the above subjects.

Anthony W. Y. LAI
BSc Thames, MSc HKPU, MRICS,
AHKIS

Anthony LAI has been qualified as a building surveyor in the Hong Kong Institute of Surveyors for nine years. His research interests, in addition to building control and maintenance, also include management and quality assurance and management.

Ellen LEE
BSc RGIT, MSc H-W, FRICS,
FHKIS, RPS

Ellen LEE had previously practiced as a quantity surveyor before she joined City University of Hong Kong. She teaches mainly quantity surveying subjects such as building measurement, building contract administration, building economics, and procurement practice. Her project experience in Hong Kong and the UK covered residential, industrial, commercial, community, fitting-out, and renovation and slope protection projects. Her research interest is varied, but focused mainly on procurement and management. Recently she has published papers on partnering and trust.

Arthur W. T. LEUNG
MPhil CityU, MCIOB, ACIArb,
MHKIE, MHKICM

Arthur LEUNG is an expert in construction management. He had worked in a wide spectrum of building projects for 13 years before he started his academic career. Specializing in construction planning, scheduling and site layout planning, he has published a number of his works in academic journals.

Kevin MANUEL
BTech Ryerson Poly., MArch NY State,
MSc(UP) HK, MBIAT, MRTPI,
Chartered Town Planner

Kevin MANUEL is involved in teaching, practicing and research as an architect-planner. He received an American Institute of Architects School Medal and a Certificate of Merit for excellence in studying architecture in 1989, while most of his architectural and planning training took place in Canada, the United States and Hong Kong. His research interests are urban design, urban renewal, public space in Hong Kong, ecological design, 20th-century China and Hong Kong architecture, and sustainable design and development.

Anna SHUM
BArch, MHKIA, Reg. Architect,
Authorized Person, M Proj Mgt,
MAIPM

Anna SHUM's professional practice experiences include working in architectural consultancy firms in Hong Kong with practical experiences in architectural and interior design, contract administration and documentation for large-scale residential and commercial construction projects both in Hong Kong and mainland China. Her research interests include project management and architectural practice.

Hong-xing WEN
BSc Tianjin (China),
MSc Tianjin (China),
MPhil Newcastle (UK),
PhD Plymouth (UK)

Hong-xing WEN has extensive teaching, consulting, and research experience in Hong Kong, mainland China and the UK in the area of structural and construction work design. He has published widely in technical reports, journals and refereed conference papers on the appraisal of reinforced concrete structures with alkali-aggregate reaction, damage and fracture mechanics modeling of concrete, and high-performance concrete containing pozzolans. His recent interests also include engineering education reform under the impact of web technology.

Joseph Francis WONG
BA(Arch) Calif., MArch MIT, MHKIA,
Reg. Architect

Joseph WONG is a member of the Hong Kong Insitute of Architects and a Registered Architect in Hong Kong. Besides teaching, he continues to practice as a consultant architect for private architectural firms on a project basis. His works have won several awards, including the Energy Efficient Building Award (1997), Second Place in the Youth Development Center (Chai Wan) design competition (2000), and Winner of the Kadoorie Pier Reconstruction (2001) design competition.

Raymond W. M. WONG
MSc CityU, Asso HKP, MCIOB,
MHKICM

Raymond WONG is a corporate member of the Chartered Institute of Building and the Hong Kong Institution of Engineers. He has undertaken studies on a great number of super-sized construction and infrastructure projects in Hong Kong. His areas of expertise and research include complex and super-high-rise building construction, super-sized underground construction, prefabricated and industrialized construction, highways, bridges and tunnel technology, as well as infrastructure and urban development studies in Hong Kong and mainland China.

Charlie Q. L. XUE
Dip(Eng), MArch, PhD TongJi ASC,
MSAA, MIAHS, MBIAT

Charlie XUE has worked in mainland China, Hong Kong, the UK and the US. He has participated in various consultancy projects in architecture as well as urban and interior design. He is the author of numerous articles in professional journals and books published in Hong Kong, mainland China, Taiwan, Europe and the US. He is also the author of two bilingual books, *Building Practice in China* (1999) and *Contemplation on Architecture* (2001).

Derek YUEN
BA(AS), BArch HK, MHKIA,
Reg. Architect

Derek YUEN is a registered architect and member of the HKIA. Before joining the academic field, he was the head of an architectural and planning department of a major property developer in Hong Kong. With over 18 years of experience in the professional field, he has worked on a wide range of town planning, infrastructure and building projects in Hong Kong, mainland China, SE Asia and Europe.

Index